Food and Free Radicals

Food and Free Radicals

Edited by

Midori Hiramatsu

Yamagata Technopolis Foundation
Yamagata, Japan

Toshikazu Yoshikawa

Kyoto Prefectural University of Medicine
Kyoto, Japan

and

Masayasu Inoue

Osaka City University Medical School
Osaka, Japan

Plenum Press • New York and London

Library of Congress Cataloging-in-Publication Data

Food and free radicals / edited by Midori Hiramatsu, Toshikazu
  Yoshikawa, and Masayasu Inoue.
       p.   cm.
    "Proceedings of the First Symposium on Food and Free Radicals,
  held June 16, 1994, in Yamagata, Japan"--T.p. verso.
    Includes bibliographical references and index.
    ISBN 0-306-45493-9
    1. Antioxidants--Congresses. 2. Food--Analysis--Congresses.
  3. Free radicals (Chemistry)--Congresses.   I. Hiramatsu, Midori.
  II. Yoshikawa, Toshikazu. III. Inoue, Masayasu. IV. Symposium on
  Food and Free Radicals (1st : 1994 : Yamagata-shi, Japan)
    [DNLM: 1. Free Radicals--metabolism--congresses. 2. Food
  Analysis--congresses. 3. Antioxidants--metabolism--congresses.
  4. Free Radical Scavengers--therapeutic use--congresses.   QD 471
  F686 1997]
  TX553.A73F63   1997
  616.07--dc21
  DNLM/DLC
  for Library of Congress   97-7267
                                                                CIP

Proceedings of the First Symposium on Food and Free Radicals, held June 16, 1994, in Yamagata, Japan

ISBN 0-306-45493-9

© 1997 Plenum Press, New York
A Division of Plenum Publishing Corporation
233 Spring Street, New York, N. Y. 10013

http://www.plenum.com

10 9 8 7 6 5 4 3 2 1

PREFACE

Natural Antioxidants and Biofactors in Food and Spice

Aerobic life utilizes huge amounts of oxygen for the maintenance of activity. Most of the inspired oxygen is used for the oxidative phosphorylation in mitochondria to generate bioenergy, while some fraction is utilized by enzymes, which directly use this gas for the metabolism of organic compounds. Furthermore, even under physiological conditions, about 3–10% of the oxygen is converted to reactive oxygen species, such as superoxide radical, hydrogen peroxide, hypochlorite, and hydroxyl radical. A generation of reactive oxygen species greatly depends on the conditions of organisms and increases in animals with inflammation and various diseases associated with circulatory disturbance. Therefore, reactive oxygen species have been postulated to play pathogenic roles in these diseases. Hence, protection of organisms from oxidative stress by inhibiting the generation of these reactive species, directly and/or by scavenging them, is believed to be of critical importance for the maintenance of aerobic life. Based on this concept, antioxidants and enzymes that scavenge reactive oxygen species have been paid much attention as therapeutic agents.

Recent studies revealed that some reactive oxygen species, such as nitric oxide (NO) and hydrogen peroxide, are also involved in the regulation of signal transduction and cellular metabolism. Therefore, excessive modulation of the metabolism of reactive oxygen species might perturb the metabolisms involved for the maintenance of physiological processes in various cells and tissues. Thus, it is important to regulate selectively the metabolism of reactive oxygen species with pathogenic nature without affecting those involved in physiological processes. Most of the hazardous oxygen species rapidly react with various components in animals. Hence, to normalize the pathologic metabolism induced by oxidative stress, antioxidant enzymes and scavengers should be targeted specifically to the site(s) of generation and/or action of reactive oxygen species. This might be particularly important for treating patients with acute inflammation and circulatory disturbance associated with oxygen stress. However, it should be noted that there is a wide variety of chronic diseases in which oxidative stress also underlies their pathogenesis, such as hypertension, atherosclerosis, diabetes, cancer, and aging. Although steady-state levels of reactive oxygen species generated in chronic diseases are fairly low, they occur continuously over a long period of time, gradually inducing cell and tissue injury, thereby increasing age-related disorders. Because potent antioxidants, which rapidly react with active oxygen species, also affect the redox states of various biomolecules involved in physiological processes, loading doses of these compounds might perturb the metabolism and dynamic balance of various compounds required for the maintenance of aerobic life. Hence, slow release and/or uptake of appropriate amounts of antioxidants and related compounds with chemically mild reactivity is of critical importance for the prevention of chronic diseases caused by oxidative stress. For this purpose, daily ingestion of sufficient amounts of antioxidants and scavengers from food is of extreme importance. Recent studies suggest that there is a wide variety of novel antioxidants and related compounds in daily food. However, physicochemical and biological properties of these beneficial compounds remain to be studied.

This volume describes the molecular mechanisms by which reactive oxygen species deteriorate various types of diseases and the structure and physicochemical properties of a wide variety of natural antioxidants involved in daily food and spices. Our knowledge about oxidative stress and antioxidants is still uncertain but is constantly improving. Hence, no one has the perfect answers at this stage, including the authors of this volume. However, this volume describes a wide variety of novel and natural antioxidants that have significant potential, not only for beneficial application in food science, but also for medical use as novel therapeutics. Before, during, and after reading this book, readers will realize the importance of natural antioxidants for a healthy life and will be reminded of the old but still important concept, "We are what we eat!"

Masayasu Inoue

CONTENTS

Part II: Antioxidant Food

Part III: Antioxidants and Diseases

Food and Free Radicals

1

FREE RADICALS IN CHEMISTRY AND BIOCHEMISTRY

Etsuo Niki

Research Center for Advanced Science and Technology
The University of Tokyo
4-6-1 Komaba, Meguro, Tokyo 153, Japan

1. PROLOGUE

100 years ago Fenton[1] found that hydrogen peroxide, although inactive alone, oxidizes tartaric acid rapidly in the presence of an iron salt. However, the real active species was not clear then and it was 40 years later that Haber and Weiss[2] proposed that the active species in Fenton reaction was the hydroxyl radical. In 1900, Gomberg[3] observed a strong experimental evidence for the formation of triphenylmethyl radical, and it is generally accepted that this is the beginning of the chemistry of free radicals. It took 30 years, however, until the chemistry of free radial chain reactions was understood. Mayo and Kharasch[4] at the University of Chicago discovered the "peroxide effect", that is, the addition of hydrogen bromide to the double bond by a radical mechanism proceeds by an Anti-Markovnikov mechanism. The decade of 1930's was an epoch of a big progress in the understanding and application of free radical chemistry. The oxidative deterioration of foods and rubber received attention in 1940's and the mechanism and kinetics of autoxidation were studied extensively since then[5]. In 1956, Harman[6] proposed a free radical theory of aging and the oxidations of hydrocarbons and their inhibition by antioxidants were studied by many groups of investigators such as Mayo, Russell and Ingold. 1969 is another important year to be commemorated since Fridovich and McCord[7] discovered superoxide dismutase (SOD), which induced a burst of research on free radicals in biology. Slater and his colleagues[8] studied free radical-induced oxidative damage of liver and Pryor[9] pointed out the importance of free radicals in biology. Since then, numerous papers have been published, which suggested that the lipid peroxidation and oxidative damage of protein and DNA are closely related with a variety of pathological events, cancer and aging[10–13]. As the experimental and epidemiological evidence suggesting the toxic effect of free radical-mediated oxidative damage in biology has been accumulated, the protective role of antioxidants has received increasing attention. Especially, the antioxidants in foods have received much attention recently in connection with their preventive function[14].

Food and Free Radicals, edited by Hiramatsu *et al.*
Plenum Press, New York, 1997

2. FREE RADICALS IN BIOLOGY

There is now ample evidence which shows the versatile role of free radicals *in vivo*. Some are positive and others negative. It has been known that superoxide anion radical plays a vital role in phagocytosis. It is also accepted that free radicals play an important role in signal transduction and induce, for example, apoptosis leading to programmed cell death. The involvement of free radicals has been often suggested in the enzymatic oxidations of fatty acids by, for example, lipoxygenase. Such oxidations are in general site and stereo-specific, which implies that the active species is not literally *free* radicals.

One of the characteristic features of the free radical reactions is that the site of radical attack is often not selective but random. The more reactive the radical is, the less selective it is. For example, hydroxyl radical attacks almost any molecule randomly. On other hand, the peroxyl radical is much more unreactive and attacks the molecule more selectively. Free radicals attack lipids, sugars, proteins and DNA to induce oxidations by a chain mechanism,

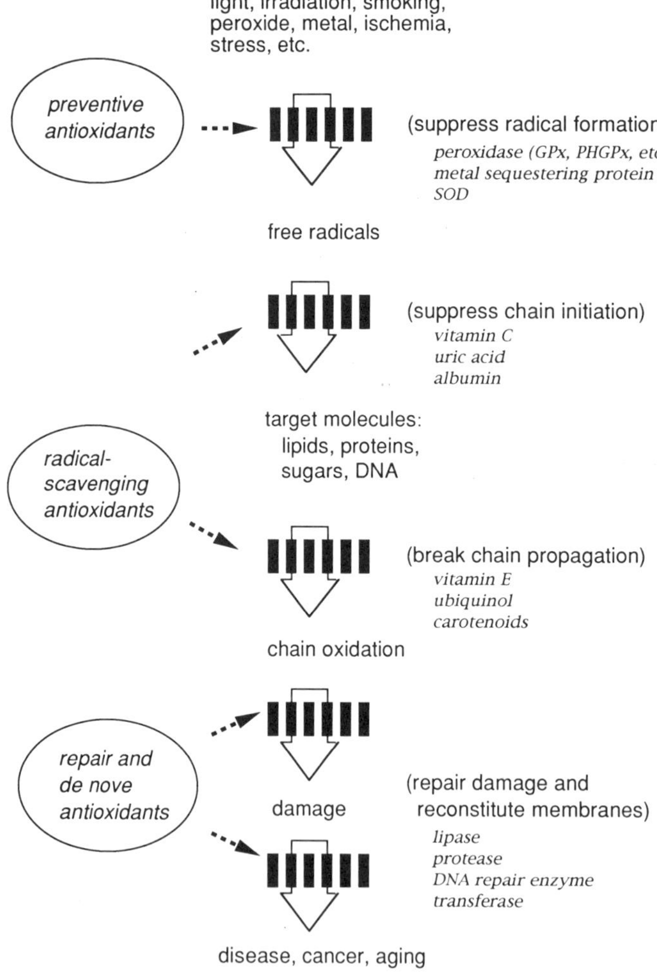

Figure 1. Defense system *in vivo* against oxidative stress.

which causes membrane damage, protein modification, enzyme deactivation, and DNA damage. They will eventually result in various diseases, cancer and aging.

We are protected against such free radical-mediated oxidative stress by an array of defense systems. As shown in Fig. 1, the preventive antioxidants such as peroxidases and metal chelating proteins suppress the generation of free radicals and act as the first line defense. The radical-scavenging antioxidants such as vitamin C and vitamin E scavenge radicals to inhibit chain initiation and break chain propagation. They are the second line defense. The repair and *de novo* enzymes act as the third line defense. For example, lipases, proteases, DNA repair enzymes and transferases repair the damage and reconstitute membranes. Furthermore, the appropriate enzymes are generated and transferred to the right site at the right time and in the right concentration by an adaptation mechanism.

It is now accepted that diet is fundamental in determining the oxidizability and also "antioxidizability" *in vivo,* that is, diet plays a critical role in determining the susceptibility to oxidative damage and efficacy of its prevention.

3. OXIDATION OF LIPIDS

The fatty acid compositions that constitute membranes and lipoproteins are determined primarily by foods. It is known that, on average, the vegetarian has less polyunsaturated fatty acids, those who eat meat have high arachidonic acid, and those who take fish have higher unsaturated acids such as eicosapentaenoic acid (EPA) and docosahexaenoic acid (DHA). Polyunsaturated fatty acids (PUFA) having two or more double bonds are quite susceptible to free radical attack and autoxidation. The relative oxidizabilities of PUFAS increase with increasing number of double bonds. In Table 1 are shown the rate constants for fatty acid oxidation[15]. It has been observed in fact that arachidonic acid is oxidized faster than linoleic acid in the oxidations of erythrocyte membranes[16] and low density lipoprotein[17].

This is because the hydrogen of the methylene groups between the 2 double bonds, called bisallylic hydrogen, is quite active toward free radical, since this C-H bond energy is small due to the resonance energy of the pentadienyl radical formed after hydrogen atom abstraction.

The bond strength of C-H bonds in PUFA is shown in Fig. 2. In SI units, bond energies are expressed in *kJ*/mol (1 cal = 4.18 J). The weakest C-H bond is 75 kcal/mol in the bisallylic position. Consequently, the bisallylic C-H bond is the most reactive site for hydrogen atom abstraction by peroxyl radicals. In the oxidation of linoleic acid and its esters, the bisallylic hydrogen at the 11-position is abstracted. However, oxygen does not

Table 1. Rate constants for reactions of fatty acids with oxygen
radicals in solution

Fatty acid	Number of double bond	k, M^{-1}s^{-1} Peroxyl LOO$^{\bullet}$	Alkoxyl RO$^{\bullet}$	Hydroxyl HO$^{\bullet}$
Stearic	0	10^{-3}~10^{-4}	2.3×10^6	10^9
Oleic	1	0.1~1	3.3×10^6	10^9
Linoleic	2	60	8.8×10^6	9×10^9
Linolenic	3	120	1.3×10^7	7×10^9
Arachidonic	4	180	2.0×10^7	10^{10}

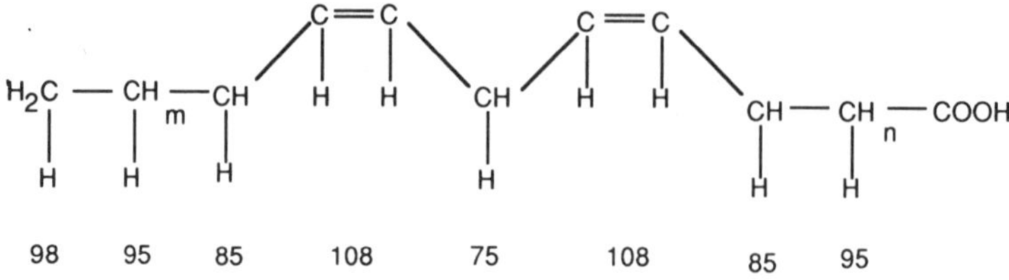

Figure 2. Bond strength of C-H bonds of polyunsaturated fatty acids, Kcal/mol.

add to the 11-position due to reduced electron density, but it adds to 9 and 13-carbon to give conjugated diene peroxyl radical[19]. The carbon-centered radicals react with oxygen at close to diffusion-controlled rates, the rate constant being over 10^9 $M^{-1}s^{-1}$. The rate constants for reaction of oxygen with PUFA radicals are smaller because of a resonant charactor of pentadienyl radical. For example, the bisallylic radical of linoleic acid reacts with oxygen with a rate constant of 1.8×10^8 $M^{-1}s^{-1}$, an order of magnitude smaller than that for saturated carbon-centered radicals but still quite high[15]. This addition reaction is reversible.

The mechanism of the free radical chain oxidation of polyunsaturated lipids depends on the number of double bond[19]. The oxidation of dienes proceeds by a scheme shown in Fig. 3 to give conjugated diene hydroperoxides quantitatively as the primary products. For example, the oxidation of linoleic acid ester gives 9-hydorperoxy-10-trans,12-cis-octadecadienoic acid, 9-hydroperoxy-10-trans,12-trans-ocatadecadienoic acid, 13-hydroperoxy-9-cis,11-trans-octadecadienoic acid, and 13-hydroperoxy-9-trans,11-trans-octadecadienoic acid esters quantitatively (Fig. 4). On the other hand, the oxidation of PUFA having more than 2 double bonds is more complicated since intramolecular addition of peroxyl radical to give cyclic peroxide competes with intermolecular hydrogen atom abstraction. For example, the oxidation of arachidonic acid and its esters gives 3 pentadienyl radicals, which react with oxygen rapidly to give 6 different peroxyl radicals (Fig. 5). 5- and 15-

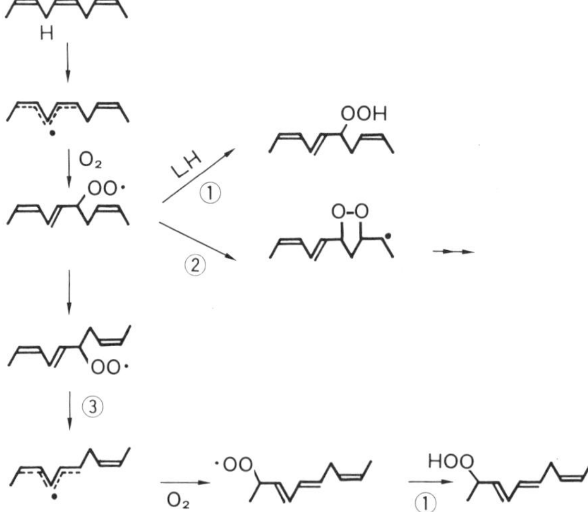

Figure 3. Oxidation mechanism of polyunsaturated fatty acids. 1, H atom abstraction; 2, intramolecular addition of peroxyl radical to double bond; 3, release of oxygen from peroxyl radical.

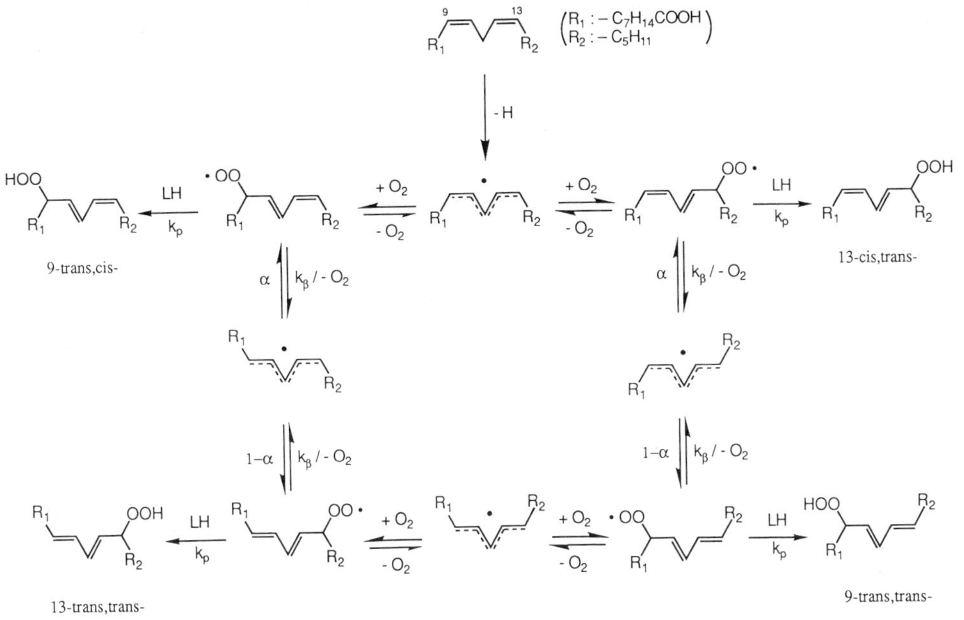

Figure 4. Oxidation of linoleic acid.

Peroxyl radicals can not add to the double bond intramolecularly and give conjugated diene hydroperoxide exclusively, whereas 8-,9-,11-,12-peroxyl radicals undergo either intramolecular hydrogen atom abstraction or intramolecular addition to give eventually prostaglandin type products. The relative importance of the 2 competing reactions is determined primarily by an availability of active hydrogen, that is, the concentration and reactivity of hydrogens. It is noteworthy that *cis,cic*-PUFA gives cis,*trans*-and *trans,trans-hydroperoxides*.

These hydroperoxides are not final stable products. They may be further oxidized to give dihydroperoxides, reduced to alcohol, and give various secondary products such as

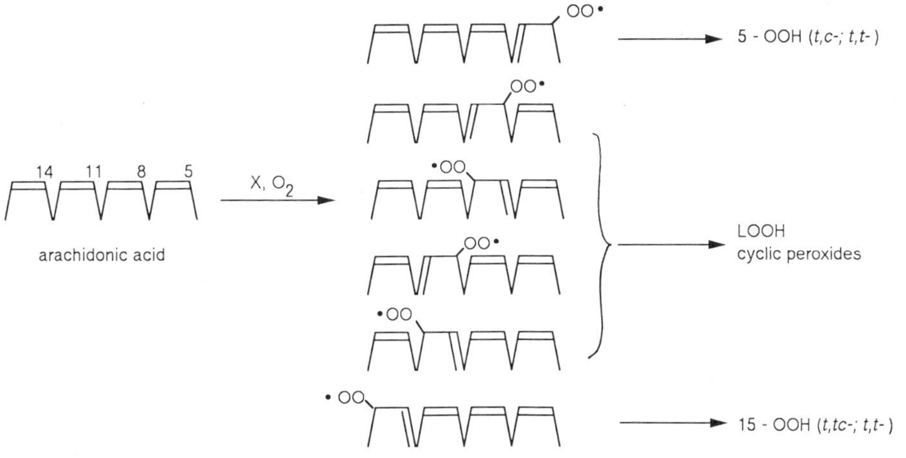

Figure 5. Arachidonic acid peroxyl radicals.

epoxides, ketones, and aldehydes. Marnett[20] observed that the alkoxyl radical gives epoxide predominantly. The formation of unsaturated aldehydes has been reported.[21] They are important since they are quite toxic.

As described above, the mechanism of oxidation of PUFA depends on the number of double bonds. However, the oxidations of PUFAs proceed by substantially the same mechanism whether they are free acids, triglycerides, phospholipids or cholesterol esters. Furthermore, they are oxidized by the same mechanism in solution, membranes and lipoproteins. The rate of oxidation may vary depending on the medium[22]. Interestingly, the relative oxidizabilities of the phosphatidylcholine and cholesterol ester were found to depend on the medium: cholesterol ester was oxidized faster than phosphatidylcholine in low density lipoprotein, but they were oxidized by the similar rate in *tert*-butyl alcohol, while phosphatidylcholine was oxidized exclusively in hexane[23].

4. OXIDATION OF PROTEINS AND DNA

Proteins are also target molecules of free radical attack. For example, all amino acid residues of a protein are subject to attack by hydroxyl radical produced by ionizing radiation, although some residues are preferred to others[24, 25]. Protein radicals formed by a free radical attack cause polypeptide chain scission, crosslinking, oxidation and modification of aminoacids. The conformational changes lead to increased susceptibility to proteolysis and heat denaturation and to loss of biological function. Under anaerobic conditions, crosslinking through -S-S- and Try-Tyr bondings is predominant, whereas in the presence of oxygen, fragmentation becomes a major reaction. The metal-catalyzed oxidations have been found to proceed site-specifically [26].

In contrast to lipid oxidation, the oxidation products from proteins have not been well characterized nor quantified yet. The formation of peroxides and charbonyl compounds have been observed[27]. Hydroxylation of the aromatic ring has been also reported[28].

It may be noteworthy that glycation of proteins by glucose to yield a Schiff base which subsequently rearranges to Amadori product may have profound effect on the susceptibilities of the proteins to oxidative damage and stress[28].

Oxidative modification of DNA has been studied extensively in radiation chemistry. Ames *et al.*[30] have estimated based on the amounts of modified purine and pyrimidine bases excreted everyday that 10^4 to 10^5 DNA bases are modified per cell per day. Several kinds of oxidation products from DNA bases have been identified.

5. PROTECTION FROM OXIDATIVE DAMAGE BY ANTIOXIDANTS

The aerobic organisms are protected from the oxidative damage induced by free radicals by a defense system outlined in Fig. 1. Numerous antioxidant enzymes and small molecules with different functions play a vital role in this defense system. Foods are important in supplying critical elements, components and compounds. Recent epidemiological studies also show that diet and risk factor for various diseases are closely related. In connection with the free radicals, two functions are required: one to suppress the formation of free radicals and the other to scavenge radicals.

Hydrogen peroxide and hydroperoxide are precursors to active free radicals and they should be detoxified. Various enzymes reduce hydrogen peroxide and hydroperoxides as

Table 2. Peroxidases

Catalase	Decomposition of hydrogen peroxide
	$2H_2O_2 \rightarrow 2H_2O+O_2$
Glutathione peroxidase (cellular)	Decomposition of hydrogen peroxide and free fatty acid hydroperoxides
	$H_2O_2+2GSH \rightarrow 2H_2O+GSSG$
	$LOOH+2GSH \rightarrow LOH+H_2O+GSSG$
Glutathione peroxidase (plasma)	Decomposition of hydrogen peroxide and phospholipid hydroperoxides
	$PLOOH+2GSH \rightarrow PLOH+H_2O+GSSG$
Phospholipid hydroperoxide glutathione peroxidase	Decomposition of phospholipid hydroperoxides
Peroxidase	Decomposition of hydrogen peroxide and lipid hydroperoxides
	$LOOH+AH_2 \rightarrow LOH+H_2O+A$
	$H_2O_2+AH_2 \rightarrow 2H_2O+A$
Glutathione-S-transferase	Decomposition of lipid hydroperoxides

summarized in Table 2. Catalase reduces hydrogen peroxide to water. Glutathione peroxidases reduces hydroperoxides to corresponding alcohols. They are selenium-containing enzymes and selenium deficiency produces a variety of diseases that are similar to those induced by vitamin E deficiency. The soil without enough selenium will cause selenium deficiency, which is accepted to be the major cause of Keshan disease.

Superoxide dismutase (SOD) is also an important antioxidant. It catalyses the dismutation of superoxide to hydrogen peroxide. Several types of SOD are known which contain copper, manganese, iron, and zinc.

Metals are essential elements. At the same time, metals especially iron and copper generate free radicals. These metals are sequestered by specific proteins *in vivo*.

Radical scavenging antioxidants are another class of important antioxidants. The antioxidant IH scavenges active radical X$^{\bullet}$ before the radical attacks target molecule.

$$X^{\bullet} + IH \xrightarrow{\ k_{inh}\ } XH + I^{\bullet}$$

It is of course necessary for the antioxidant to scavenge the radical rapidly, that is, the antioxidant must be reactive toward the radical and the rate constant k_{inh} should be large. Another important factor is the fate of antioxidant-derived radical I$^{\bullet}$. It may dimerize by reacting with another I$^{\bullet}$ and scavenge another X$^{\bullet}$ radical to give an adduct. It may be also reduced by some reductant to regenerate the antioxidant. If the I$^{\bullet}$ radical is not stable enough, it may react with the target molecule to induce radical-mediated oxidation. This will diminish the antioxidant activity.

Among the many radical-scavenging antioxidants, the chemistry of inhibition of oxidation has been studied extensively for α-tocopherol, the most biologically active form of vitamin E. α-Tocopherol reacts rapidly with hydroxyl, alkoxyl and peroxyl radicals, but the scavenging of peroxyl radical must be the principal role of α-tocopherol as an antioxidant, since many substrates react with hydroxyl and alkoxyl radicals at similar rates. When α-tocopherol scavenges radicals, α-tocopheroxyl radical is formed. As summarized in Fig. 6, it may react with another α-tocopheroxyl radical to give a dimer or scavenge another peroxyl radical to give an adduct. These products have been isolated and confirmed[31–33]. The peroxyl radical adduct may undergo secondary reactions further to give α-tocopheryl quinone eventually.

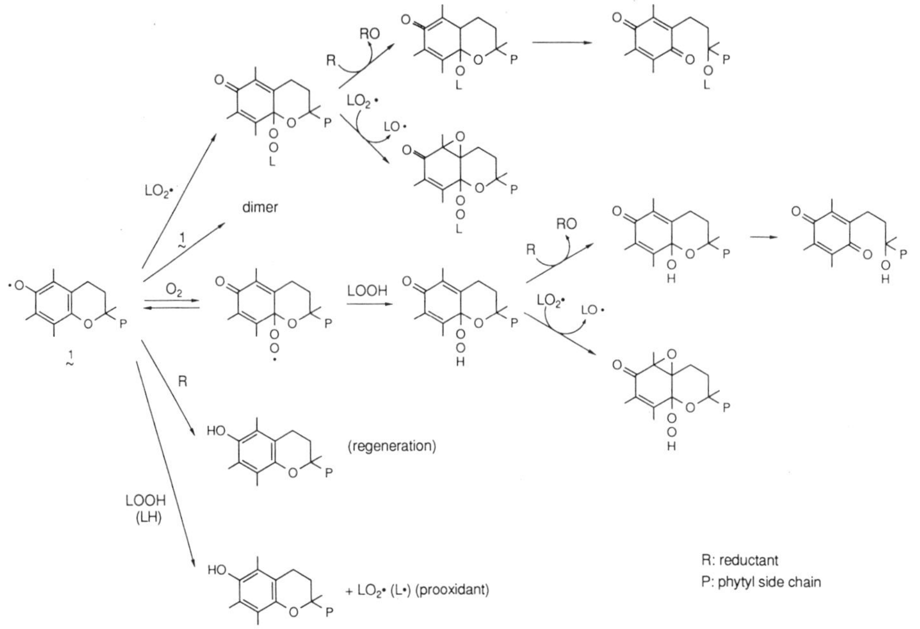

Figure 6. Fate of vitamin E radical.

α-Tocopheroxyl radical is reduced rapidly by ascorbate and ubiquinol to regenerate α-tocopherol[34]. Ascorbate and ubiquinol are potent antioxidants *per se*, but the reduction of α-tocopheroxyl radical may be another important role. It has been known that α-tocopherol accelerates the oxidation of lipids under some conditions[35]. Such dual function or crossover effect is observed especially at its high concentrations[36, 37]. Similar effect has been reported in the oxidation of low density lipoprotein recently[35]. It may be noteworthy that antioxidant is also a double-edged sword like oxygen.

The beneficial role of vitamin E has been reported by many investigators[38,39]. For example, the recent epidemiological studies show that vitamin E is effective in reducing the risk of coronary heart disease[40,41]. The classic diet-heart hypothesis states that populations with high intakes of saturated fat will have high rates of coronary heart disease. It is also known that some population groups in Europe have relatively low rates of coronary heart disease despite high saturated fat intakes. This phenomenon, referred to as the "French paradox" has been explained by the polyphenolic antioxidants in the red wine[42,43]. It has been pointed out that dietary vitamin E may provide at least as good an explanation as does the red wine for "French paradox"[44].

In addition to vitamin E many other radical-scavenging antioxidants have been found or synthesized and their antioxidant potency has been studied. Among others, phenolic compounds, polyphenols and flavonoids found and isolated from plants, vegetables and fruits have received much attention. They will be covered in other chapters of this volume.

6. EPILOGUE

The chemistry of free radicals started at the beginning of this century. Free radical-mediated reactions such as oxidation and polymerization have been applied successfully in the modern chemical industries. At the same time, free radicals also caused negative effects such as air pollution and oxidative degradation and deterioration of rubber, plastics, oils and foods. There is now an ample evidence that free radicals act similarly *in vivo*. They play an important role in the synthesis of biologically active molecules, phagocytosis and probably signal transduction. The oxidative stress is the adverse effect of free radicals. It is essential to control the reactions by promoting the positive reactions and suppressing the negative ones. This will be the target of our future studies.

REFERENCES

1. Fenton, H. J. H. (1894) Oxidation of tartaric acid in the presence of iron. J. Chem. Soc. 65: 899–910.
2. Haber, F. and Weiss, J. J. (1934) The catalytic decomposition of H_2O_2 by iron salts. Proc. Royal Soc.London, Biol. A147:332–351.
3. Gomberg, M. (1900) An instance of trivalent carbon: Triphenylmethyl. J. Am. Chem. Soc. 22: 757–771.
4. Kharasch, M. S. and Mayo, F. R. (1933) The peroxide effect in the addition of reagents to unsaturated compounds. I. The addition of hydrogen bromide to allyl bromide. J. Am. Chem. Soc. 55: 2468–2496.
5. Bateman, L. and Gee, G. (1949) The determination of absolute rate constants in olefinic oxidations. Proc. Roy. Soc. A195: 391–402.
6. Harman, D. (1956) Aging: a theory based on free radical and radiation chemistry. J. Gerontol. 11: 298–300.
7. McCord, J. M. and Fridovich, I. (1969) An enzymic function for erythrocuprein (hemocuprein). J. Biol. Chem. 244: 6049–6055.
8. Slater, T. F. (1968) Biochem. J. 106: 155.
9. Pryor, W. A. (1970) Free radicals in biological systems. Scientific Americans 223: 70–83.
10. Pryor, W. A. ed. (1976–1984) Free Radicals in Biology. Vol. I-VI. Academic Press, Orlando.
11. Yagi, K. ed. (1982) Lipid Peroxide in Biology and Medicine, Academic Press, New York.
12. Halliwell, B. and Gutteridge, O. (1985) Free Radicals in Biology and Medicine, Clarendon Press, Oxford.
13. Sies, H. (1985) Oxidative Stress, Academic Press, London.
14. Slater, T. F. and Block, G. (1991) Antioxidant Vitamins and β-Carotene in Disease Prevention, Am. J. Clin. Nutr. 53, Suppl.
15. Simic, M. G., Jovanovic, S. V. and Niki, E. (1992) Mechanisms of lipid oxidative processes and their inhibition. In: A. J. St. Angelo (ed), Lipid Oxidation in Food. ACS Symp. Ser. 500, pp.14–32, Am. Chem. Soc., Washington, D. C.
16. Miki, M., Tamai, H., Mino, M., Yamamoto, Y. and Niki, E. (1987) Arch. Biochem. Biophys.258: 373–380.
17. Quehenberger, O., Koller, E., Jurgens, G. and Esterbauer, H. (1987) Investigation of lipid peroxidation in human low density lipoprotein. Free Rad. Res. Comms. 3: 233–242.
18. Simic, M. G., Jovanovic, S. V. and Al-Sheikhly, M. (1989) Heterocyclic resonant radicals. Free Rad. Res. Comms. 6: 113–115.
19. Porter, N. A., Lehman, L. S., Weber, B. A. and Smith, K. J. (1981) Unified mechanism for polyunsaturated fatty acid autoxidation. Competition of peroxy radical hydrogen atom abstraction, β-scission, and cyclization. J. Am. Chem. Soc. 103: 6447–6455.
20. Wilcox, A. L. and Marnett, L. J. (1993) Polyunsaturated fatty acid alkoxyl radicals exist as carbon-centered epoxyallylic radicals: A key step in hydroperoxide-amplified lipid peroxidation. Chem. Res. Toxicol. 6: 413–416.
21. Esterbauer, H., Schaur, R. J. and Zollner, H. (1991) Chemistry and biochemistry of 4-hydroxynonenal, malonaldehyde and related aldehydes. Free Rad. Biol. Med. 11: 81–128.
22. Barclay, L. R. C. (1993) Model biomembranes: quantitative studies of peroxidation, antioxidant action, partitioning, and oxidative stress. Can. J. Chem. 71: 1–16.
23. Noguchi, N., Gotoh, N. and Niki, E. (1993) Dynamics of the oxidation of low density lipoprotein induced by free radicals. Biochim. Biophys. Acta 1168: 348–357.
24. Stadtman, E. R. (1993) Oxidation of free amino acids and amino acid residues in proteins by radiolysis and by metal-catalyzed reactions. Annu. Rev. Biochem. 62: 797–821.
25. Dean, R. T. (1987) Free radicals, membrane damage and cell-mediated cytolysis. Br. J. Cancer 55: Suppl. VIII:39–45.
26. Uchida, K. and Kawakishi, S. (1990) Site-specific oxidation of angiotensin I by copper (II) and L-ascorbate; conversion of histidine residues to 2-imidazolones. Arch. Biochem. Biophys. 283: 20–26.

27. Gebicki, S. and Gebicki, J. M. (1993) Formation of peroxides in amino acids and proteins exposed to oxygen free radicals. Biochem. J. 289: 743–749.

28. Simpson, J. A., Narita, S., Gieseg, S., Gebicki, S., Gebicki, J. M. and Dean, R. T. (1992) Long-lived reactive species on free-radical-damaged proteins. Biochem. J. 282: 621–624.

29. Wolff, S. P., Jiang, Z. J. and Hunt, J. V. (1991) Protein glycation and oxidative stress in diabetes mellitus and aging. Free Rad. Biol. Med. 10: 339–352.

30. Ames, B. N., Shigenaga, M. K. and Hagen T. M. (1993) Oxidants, antioxidants, and the degenerative diseases of aging. Proc. Natl. Acad. Sci. USA 90: 7915–7922.

31. Matsuo, M., Matsumoto, S., Iitaka, Y. and Niki, E. (1989) Radical scavenging reactions of vitamin E and its model compound, 2,2,5,7,8-pentamethyl-6-chromanol, in a *tert*-butylperoxyl radical-generating system. J. Am. Chem. Soc. 111: 7179–7185.

32. Yamauchi, R., Yagi, Y. and Kato, K. (1994) Isolation and characterization of addition products of α-tocopherol with peroxyl radicals of dilinoleoylphosphatidylcholine in liposomes. Biochim Biophys Acta 1212: 43–49.

33. Liebler, D. C., Baker, P. F. and Kaysen, K. L. (1990) Oxidation of vitamin E: Evidence for competing autoxidation and peroxyl radical trapping reactions of the tocopheroxyl radical. J. Am. Chem. Soc. 112: 6995–7000.

34. McCay, P. B. (1985) VITAMIN E: Interactions with free radicals and ascorbate. Ann. Rev. Nutr. 5: 323–340.

35. Bowry, V. W., Ingold, K. U. and Stocker, R. (1992) Vitamin E in human low-density lipoprotein. Biochem. J. 288: 341–344.

36. Terao, J. and Matsushita, S. (1986) The peroxidizing effect of α-tocopherol on autoxidation of methyl linoleate in bulk phase. Lipids 21: 255–260.

37. Takahashi, M., Yoshikawa, Y. and Niki, E. (1989) Crossover effect of tocopherol in the spontaneous oxidation of methyl linoleate. Bull. Chem. Soc. Jpn. 62: 1885–1890.

38. Sies, H. (1993) Efficacy of vitamin E in the human. VERIS News.

39. Chow, C. K. (1991) Vitamin E and oxidative stress. Free Rad. Biol. Med. 11: 215–232.

40. Stampfer, M. J., Hennekens, C. H., Manson, J. E., Colditz, G. A., Rosner, B. and Willett, W. C. (1993) Vitamin E consumption and the risk of coronary disease in women. New Eng. J. Med. 328:1444–1449.

41. Rimm, E. B., Stampfer, M. J., Ascherio, A., Govaninucci, E. , Coldits, G. A. and Willett, W. C. (1993) Vitamin E consumption and the risk of coronary heart disease in men. New Eng. J. Med. 328: 1450–1456.

42. Rimm, E. B., Giovannucci, E. L., Willett, W. C. Goldits, G. A. (1991) Prospective study of alcohol consumption and risk of coronary disease in men. Lancet 338: 464–468.

43. Kondo, K., Matsumoto, A., Kurata, H., Takahashi, H., Koda, H., Amachi, T. and Itakura, H. (1994) Inhibition of oxidation of low density lipoprotein with red wine. Lancet 344: 1152.

44. Bellizzi, M. C., Franklin, M. F., Duthie, G. G. and James W. P. T. (1994) Vitamin E and coronary heart disease: the european paradox. Europ. J. Clin. Nutr. 48: 822–831.

2

FREE RADICALS AND DISEASES

Toshikazu Yoshikawa, Yuji Naito, and Motoharu Kondo

First Department of Medicine
Kyoto Prefectural University of Medicine
Kamigyo-ku, Kyoto 602, Japan

1. INTRODUCTION

In recent years, it has been shown that reactive oxygen species or free radicals are closely involved in various biological reactions. At normal rate of generation, some free radicals are useful in the human body. However, when oxygen free radical generation exceeds the capacity of antioxidant defenses, the result is oxidative stress. Oxidative stress occurs in many human diseases and sometimes makes a significant contribution to their pathogenesis. Especially, oxygen free radicals such as superoxide anion radical and hydroxyl radical, are thought to mediate a large portion of the tissue damage produced after inflammation, ischemia, and ischemia/reperfusion of the small intestine, stomach, heart, kidney, liver and skin, and may also be involved in the pathogenesis of circulatory shock and disseminated intravascular coagulation. In addition, reactive oxygen metabolites and free radicals are involved in many pathogenic conditions via DNA damage, inactivation of nitric oxide (NO)(1), oxidation of LDL, and activation of adhesion molecules(2–4) (Figure 1). Herein, we summarized recent data that supported free radical involvement in ischemia/reperfusion injury and digestive diseases, and reviewed the antioxidant therapy for these pathophysiological conditions.

2. FREE RADICALS AND LIPID PEROXIDATION

By definition, a free radical is a molecule or atom with an unpaired electron. By virtue of their unpaired electron, free radicals are usually unstable and quite reactive. Highly reactive oxygen radicals such as hydroxyl radical have half lives in the nanosecond to millisecond range. In normal oxidative phosphorylation molecular oxygen is reduced by 4 electrons to form H_2O by the following sequence (Figure 2):

$$O_2 \xrightarrow{e^-} O_2^{\bullet-} \xrightarrow{e^-} H_2O_2 \xrightarrow{e^-} {}^{\bullet}OH \xrightarrow{e^-} H_2O \tag{1}$$

Food and Free Radicals, edited by Hiramatsu *et al.*
Plenum Press, New York, 1997

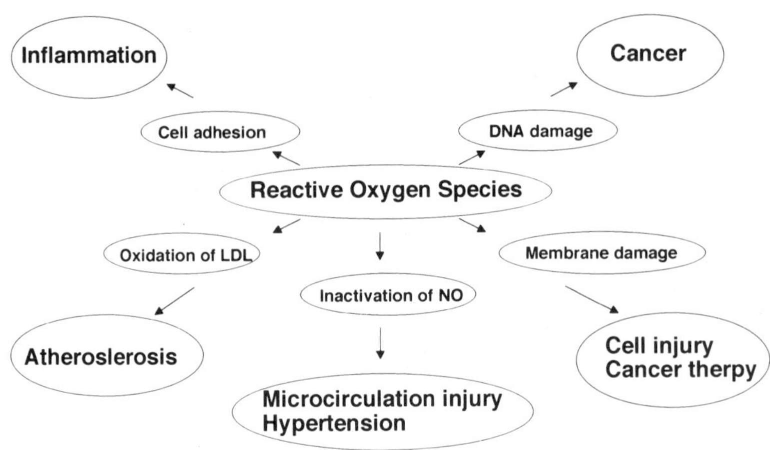

Figure 1. Role of reactive oxygen species in various pathological conditions.

Both superoxide anion ($O_2^{\cdot-}$) and hydroxyl radical ($^{\cdot}OH$) are free radicals which have the potential reacting with biological macromolecules and thereby inducing tissue damage. Hydrogen peroxide (H_2O_2) is itself a less potent oxidizing agent but in the presence of transition metal ions such as iron, $O_2^{\cdot-}$ converts ferric to ferrous, which then can react with H_2O_2 to generate more reactive hydroxyl radicals ($^{\cdot}OH$). Since most molecules formed under physiologic conditions do not have unpaired electrons, free radicals abstract an electron from a stable compound that, in turn, is transformed into a new free radical. Therefore, free radical reactions tend to proceed as chain reactions. This chain reaction will continue until the free radical is deactivated by a chain reaction-breaking antioxidant. The most-studied free radical chain reaction in living systems is lipid peroxidation, which is mediated by oxygen free radicals and is believed to be an important cause of cell membrane destruction and cell damage.

Lipid, nucleic acids, enzymes, and protein are important target molecules of biological damage caused by oxygen free radicals. In particular, polyunsaturated fatty acids (PUFAs) located in the lipophilic section of the cell membranes are prone to be attacked by oxygen radicals that produce lipid peroxides through a chain reaction of lipid peroxidation (Figure 3).

$O_2^{\cdot-}$	superoxide radical
H_2O_2	hydrogen peroxide
$HO^{\cdot}$	hydroxyl radical
$^1O_2$	singlet oxygen
$HOO^{\cdot}$	hydroperoxyl radical
$ROOH$	alkylhydroperoxide
$ROO^{\cdot}$	alkylperoxyl radical
$RO^{\cdot}$	alkoxyl radical
ClO^-	hypochlorite ion
$Fe^{4+}O$	ferryl ion
$Fe^{5+}O$	periferryl ion
$NO^{\cdot}$	nitric oxide

Figure 2. Reactive oxygen species and free radicals.

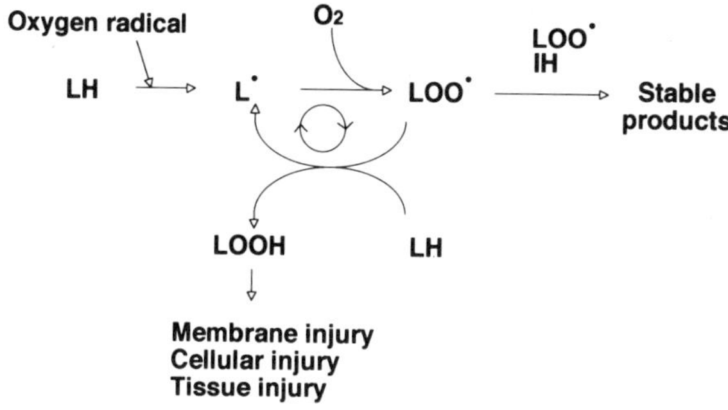

Figure 3. Free radical chain reaction.

PUFAs contain two or more carbon-double bonds within their structure. Peroxidation is initiated by the attack of any chemical species that has sufficient reactivity to abstract a hydrogen atom from a methylene carbon in unsaturated fatty acid, including hydroxyl radical and metal ion-free radical complexes. As shown in Figure 3, the resulting carbon-centered radical(L˙) undergoes molecular rearrangement, followed by reaction with oxygen to give a peroxyl radical (LOO˙). The lipid peroxyl radical attacks another lipid molecule and abstracts hydrogen to give lipid hydroperoxide and at the same time another lipid radical. The lipid radical starts the propagation sequence over again (5). When generated in membranes, peroxyl radicals can react with amino acid residues on membrane proteins, impairing the functions of proteins. They can abstract hydrogen atoms from adjacent PUFA side chains, thus propagating the free radical chain reaction of lipid peroxidation. Therefore, a single initiating event can result in conversion of hundreds of fatty acid side chains into lipid peroxides, which alter the structural integrity and biochemical functions of membranes. In the presence of iron and copper complexes, lipid peroxides decompose to form peroxyl and alkoxyl radicals, which have sufficient reactivity to initiate further peroxidation of lipids. These pathological changes may attack the endothelial cells and parenchymals cells of the digestive tract under several pathological conditions, such as acute gastric mucosal injury, ischemic colitis, radiation colitis, inflammatory bowel diseases, drug-induced liver damage, and acute pancreatitis. Therefore, it is important to enhance the activity of the endogenous antioxidants or to administer synthetic exogenous antioxidants, in order to prevent organs from the injury mediated by lipid peroxidation.

3. OXYGEN RADICAL PRODUCTION BY INFLAMMATORY CELLS

Inflammatory cells like eosinophils, neutrophils, monocytes, and macrophages become activated during inflammation and ischemia/reperfusion. The potential source of oxygen radicals in postischemic tissues and inflammatory gastro-intestinal mucosa is the activated neutrophils(6, 7). Oxygen-derived free radicals generated from eosinophils play a critical role in asthma and are closely linked with bronchial hyperresponsiveness. Due to this activation, the respiratory burst occurs, that is, the cells take up oxygen and produce the superoxide anion radicals (Figure 4). The reduction of oxygen to the superoxide anion radical is catalyzed by the NADPH oxidase enzyme. The produced superoxide anion radical dismutases to hydrogen peroxide either spontaneously (with a rate constant of 10^5 $M^{-1}s^{-1}$) or catalyzed by superoxide

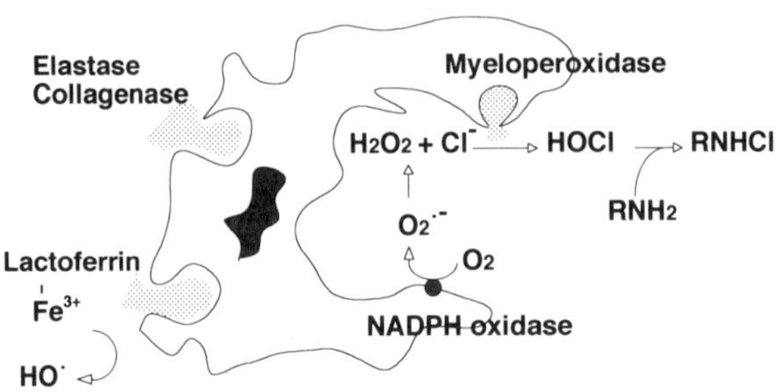

Figure 4. Oxygen radical production by phagocytes.

dismutase at a higher rate (10^9 $M^{-1}s^{-1}$). Hydrogen peroxide is a potent yet sluggish oxidant which can enter cells and consequently react with a catalytically active form of iron (ferritin) to produce the highly reactive and cytotoxic hydroxyl radical ($\cdot$OH). Ferric ions in physiologically conditions chelated by ferritin can be reduced by superoxide anion radicals resulting in release of the metal from the protein as a ferrous ion. Free ferrous ions may participate in the Fenton reaction, producing hydroxyl radicals.

Hydrogen peroxides formation in proximity of chloride ions, abundant in gastrointestinal secretions, and certain peroxidases, especially myeloperoxidase (MPO) secreted into the extracellular medium by activated leukocytes, result in the formation of the hypochlorous acid HOCl, a potent oxidant that is the active ingredient in household bleach. HOCl is well-recognized oxidizing and chlorinating agent that reacts with promary amins (RNH_2) to yield N-chloro derivatives (chloramine, RNHCl). Chloramines are known to cause cytotocxicity through sulfhydryl oxidation, cytochrome inactivation, chlorination of purine bases on DNA, and degradation of proteins and amino acids(8). In addition, monochloramine is an extremely effective promotor of increased vascular permeability, electrolyte loss, and epithelial cell loss in the colon.

Recent studies have reported that Helicobacter pylori plays an important role in the pathogenesis of active chronic gastritis and peptic ulcers(9, 10). H. pylori is characterized by high urease activity, which metabolizes urea to ammonia. H. pylori-associated chronic active gastritis is characterized by an invasion of neutrophils in the gastric mucosa. Although the mechanism by which H.pylori-induced lesion is produced is unclear, Suzuki et al.(11) has shown that H.pylori-activated neutrophils promote gastric mucosal cell injury and that monochloamine play a crucial role in this process. Recent report show that a water extract of H.pylori promotes leukocyte adhesion and emigration in mesenteric venules via CD11a/CD18- and CD11b/CD18-dependent interaction with ICAM-1(12). Further clinical studies will be necessary to investigate the efficacy of free radical scavenger or anti-adhesion therapy for the treatment of H.pylori-associated gastric lesions.

4. ISCHEMIA-REPERFUSION INJURY

There is a growing body of evidence indicating neutrophils contribute to the microvascular dysfunction associated with reperfusion of ischemic tissues. Ischemia-reperfusion results in the rapid recruitment of circulating neutrophils in gastrointestinal tissues. The interactions between neutrophils and endothelial cells are mediated by a variety of immunospecific adherence receptor-ligand pairs that are coordinately regulated on neutro-

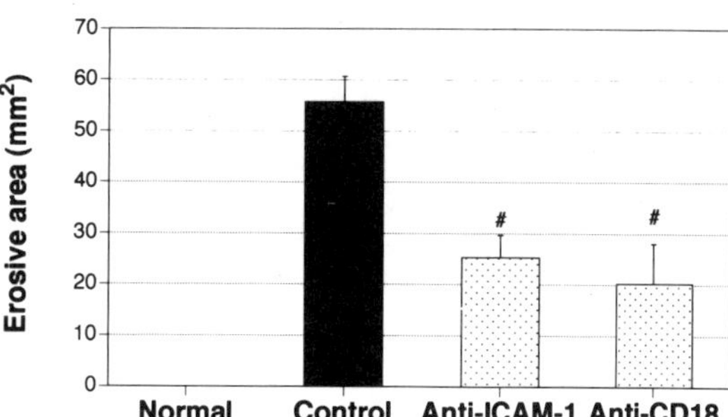

Figure 5. Effects of anti-ICAM-1 and anti-CD18 antibody on acute gastric mucosal injury induced by ioschemia-reperfusion in rats. #p<0.05 compared with the control.

phils and endothelial cells. Among these, the CD11/CD18 glycoprotein complex expressed by activated neutrophils has been shown to play an important role in neutrophil adherence to endothelial cells in a variety of in vivo and in vitro models of inflammation and ischemia-reperfusion. The CD11/CD18 complex consists of three heterodimers, each of which is comprised of an immunologically distinct α-subunit (CD11) and a common β_2-subunit (CD18), and they are classified according to their unique α-subunit, i.e. CD11a/CD18, CD11b/CD18, and CD11c/CD18. CD11a/CD18 and CD11b/CD18 are generally considered the most important heterodimers involved in neutrophil-endothelial cell interactions. Neutrophil-endothelial cell interactions are facilitated by several immunospecific adherence receptor-ligand pairs that are coordinately regulated on neutrophils and endothelial cells. Among these, CD11/CD18 glycoproteins have been shown to play a major role in both neutrophil adhesions to endothelial cells and neutrophil transendothelial migration in vitro and in vivo. Yoshida et al.(3) have reported that anoxia /reoxygenation promotes neutrophil adherence via CD11a/CD18- and CD11b/CD18-dependent interactions with ICAM-1 that is mediated by hydrogen peroxides and platelet activating factor. We have proved that the immunoneutralization of the CD11/CD18 adhesion complex on neutrophils by anti-CD18 MAb or of the ICAM on endothelial cells by anti-ICAM-1 antibody reduces both neutrophil accumulation in the gastric mucosa and gastric mucosal damage induced by ischemia-reperfusion in rats (Figure 5).

5. SUPEROXIDE AND NITRIC OXIDE

In 1980, Furchgott and Zawadzki(13) showed that the presence of endothelial cells was obligatory for the acetylcholine-induced relaxation of the rabbit aorta. Subsequent work has made it clear that the endothelium can synthesize and release not only vasodilator substances, such as prostacyclin and endothelium-derived relaxing factor (EDRF), but also vasoconstrictor substances, such as endothelin, superoxide radical and endothelium-derived contracting factors (EDCFs). The major pharmacological properties of EDRF are vascular smooth muscle relaxation and inhibition of platelet aggregation. In addition, an EDRF plays a vital role in cell communication, neutrophil activity, and perhaps ischemia and other pathological conditions. It is thought that EDRF may be nitric oxide (NO), which is a superoxide scavenger and by that protects the endothelium from free radicals released from the circulating leukocytes. Superoxide dismutase (SOD) enhances the relaxant effects of NO, supporting the view that NO is readily inactivated by superoxide via an

Table 1. Effects of NO and L-arginine related compounds on acute gastric mucosal injury induced by ischemia-reperfusion in rats

Group	(n)	(mm^2)
Sham operation	7	0.0+0.0
Ischemia-Reperfusion		
+ saline (i.v.)	10	49.2+9.3
+ SNP (6 µg/kg/min, i.v.)	7	33.4+8.3
+ SNP (12 µg/kg/min, i.v.)	7	20.0+7.8*
+ D-arginine (300 mg/kg, i.v.)	9	54.7+5.2
+ LNNA (10 mg/kg, i.v.)	10	84.6+6.2**
+ LNNA + L-arginine (300 mg/kg, i.v.)	10	57.7+8.4

Each value indicates the mean+SE.
SNP: sodium nitroprusside, LNNA: NG-nitro-L-arginine
*p<0.05 vs. saline group, **p<0.05 vs. D-arginine group.

intermediate formation of peroxynitrite (reaction 1) and subsequent conversion to nitrate(14).

$$NO + O_2^- \rightarrow ONOO^-$$ (2)

It has been suggested that generation of superoxide radicals at the time of reperfusion may be responsible for reperfusion injury in the heart, stomach, intestine, and kidney. NO may be exhibit protective activity under these conditions due to its ability to scavenge superoxide. Our recent study has proved that N^G-nitro-L-arginine (LNNA), a specific inhibitor of NO synthase, significantly promotes the gastric mucosal injury induced by ischemia-reperfusion in rats, and that sodium nitroprusside (SNP), an exogenous NO Donner, significantly inhibited this injury (Table 1). However, the peroxynitrate intermediate formed by reaction between superoxide and NO may yield hydroxyl radical upon breakdown to NO_2 (reaction 2)(1).

$$ONOO^- + H^+ \rightarrow NO_2^{\bullet} + {}^{\bullet}OH$$ (3)

If this occurs in vivo as well, the potential toxicity of superoxide may not be mitigated at all by its reaction with NO. Further investigation is necessary for the role of the interaction between superoxide and NO in some pathological conditions.

6. ANTIOXIDANT THERAPY IN DIGESTIVE DISEASES

The antioxidant defense system is divided into two parts : the preventive antioxidants and chain breaking antioxidants. From this classification, we have examined the effect of antioxidant therapy in digestive diseases and evaluated the antioxidant activity of a novel synthetic compound. Vitamin E functions as a potent chain-breaking lipid -soluble antioxidant in the living system by scavenging oxygen radicals and terminating free radical chain reaction. In an ischemia-reperfusion injury model in rats, the level of alpha-tocopherol in the gastric mucosa tended to decrease during ischemia and decreased further during early reperfusion(15). The alpha-tocopherol/cholesterol ratio also decreased after ischemia-reperfusion. These findings suggest that vitamin E may be consumed in the serum and the gastric mucosa to prevent the development of tissue damage. We studied the degree of gastric mucosal injury induced by ischemia-reperfusion in rats raised on a vitamin E-deficient (total tocopherols content: <0.1 mg/100 g diet) , -sufficient diet (2.0 I.U. of Dl-α-tocopheryl acetate/100 g diet.), and -supplemented diets (20.0 I.U. of Dl-α-tocopheryl acetate/100 g diet). Vitamin E-deficiency significantly aggravated this gastric mucosal injury and also significantly increased lipid peroxides in the gastric mucosa compared with the vitamin E-sufficient rats. These results show that cellular damage caused by gastric ischemia and reperfusion can be explained by free radical reaction processes during ischemia and reperfusion, and that vitamin E plays an important role as an endogenous antioxidant to protect tissues from oxidative stress during ischemia-reperfusion.

Recently, a new therapeutic approach using agents with antioxidant properties has been proposed. Polaprezinc, N-(3-aminopropionyl)-L-histidinato zinc, is a chelate compound consisting of zinc ion and L-carnosine, which attenuates the gastric mucosal injuries and the increase in lipid peroxides of the gastric mucosa induced by burn shock and ischemia-reperfusion in rats without any effects on gastric mucosal blood flow, acid secretion, or prostagladins<Yoshikawa, 1989 #64; Yoshikawa, 1991 #63>. This agent also inhibits the aggravation of acute pancreatitis induced by cerulein. In vitro studies have shown that polaprezinc scavenges superoxide anion and singlet oxygen, and inhibits the superoxide production by polymorphonuclear leukocytes and the generation of hydroxyl radicals by Fenton reaction as well(16). These findings indicate that the strong anti-ulcer and antioxidative actions of polaprezinc in vivo are dut to a combination of these antioxidant actions in vitro.

We have reported that several kinds of drugs, traditional Kampo medicines, and natural nutrients used to treat to humans with several disease and to maintain our healthy condition, have some antioxidant activities in vitro and in vivo. A Kampo agent, TJ-35(Shigyaku-san), significantly inhibited the increases in gastric mucosal erosions (Figure 6) and in lipid peroxide of the gastric mucosa after ischemia-reperfusion in rats, and demonstrated superoxide radical- and hydroxyl radical-scavenging activities in vitro(17). Catechin is one of the flavonoids and rich in various plants, vegetables, and teas. Green tea extract (GTE), including epicatechin, epicatechin gallate, epigalocatechin, and epigalocatechin, significantly decreased ischemia/reperfusion-induced gastric mucosal injury to 39.8 ± 18.1 mm^2 from 77.8 ± 11.6 mm^2 (control). The ESR spin trapping study has proved

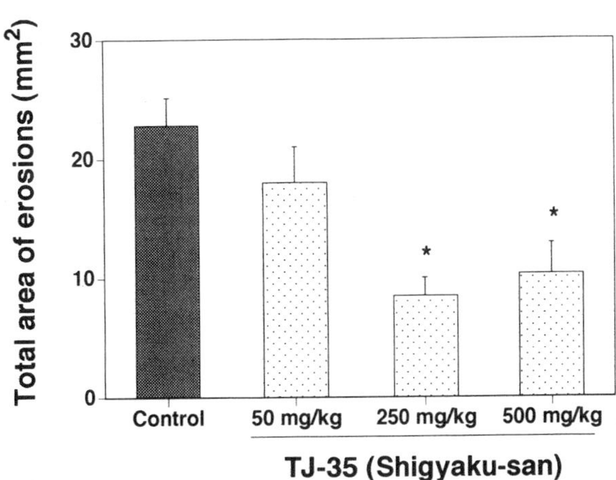

Figure 6. Effect of TJ-35(Shigyaku-san) on acute gastric mucosal injury induced by ischemia-reperfusion in rats. *p<0.05 compared with the control.

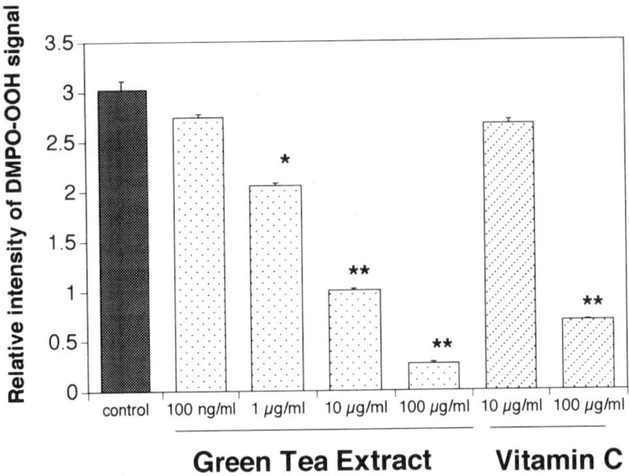

Figure 7. Superoxide scavenging activity of green tea extract compared with that of vitamin C determined by the EPR spin trapping method. *p<0.05, **p<0.01 compared with the control.

that GTE scavenges superoxide radical, hydroxyl radical, and DPPH radical in a concentration-dependent manner (Figure 7). In addition, GTE inhibited lipid peroxidation in rats brain homogenate and in gastric mucosal homogenate. These results show that the gastroprotective action of GTE against ischemia-reperfusion injury may be attributed to its antioxidative activities.

7. CONCLUSIONS

We have summarized and reviewed free radical mechanisms leading to patogenicity of human diseases. Several potential therapeutics with antioxidant properties are currently being tested. Scavenging of oxygen radicals may be beneficial to the medical status of patients with some digestive diseases.

REFERENCES

1. Beckman, J. S., Beckman, T. W., Chen, J. et al.(1990) Apparent hydroxyl radical production by peroxynitrite : Implications for endothelial injury from nitric oxide and superoxide. Proc. Natl. Acad. Sci. USA 87: 1620–1624.
2. Patel, K.D., Zimmerman ,G.A., Prescott, S.M. et al. (1991) Oxygen radicals induce human endothelial cells to express GMP-140 and bind neutrophils. J. Cell Biol. 112: 749–759.
3. Yoshida, N., Granger, D.N., Anderson, D.C. et al. (1992) Anoxia/reoxygenation-induced neutrophil adherence to cultured endothelial cells. Am. J. Physiol. 262: H1891-H1898.
4. Davenpeck, K.L., Gauthier, T.W. and Lefer, A.M.(1994) Inhibition of endothelial-derived nitric oxide promotes P-selectin expression and actions in the rat microcirculation [see comments]. Gastroenterology 107: 1050–1058.
5. Niki, E.(1987) Antioxidants in relation to lipid peroxidation. Chem. Phys Lipids 44: 227–253.
6. Smith, S.M., Holm-Rutili, L., Perry, M.A. et al.(1987) Role of neutrophils in hemorrhagic shock-induced gastric mucosal injury in the rat. Gastroenterology 93: 466–471.
7. Granger, D.N. (1988) Role of xanthine oxidase and granulocytes in ischemia-reperfusion injury. Am. J. Physiol. 255: H1269-H1275.
8. Grisham, M.B. and Granger, D.N.(1988) Neutrophil-mediated mucosal injury -role of reactive oxygen metabolites. Diges. Dis. Science 33: 6S-15S.

9. Warren, J.R. and Marshall, B.(1983) Unidentified curved bacilli on gastric epithelium in active chronic gastritis. Lancet I: 1273–1275.

10. Marshall, B.J. (1991) Virulence and pathogenicity of Helicobacter pylori, J. Gastroenterol. Hepatol. 8:121–124.

11. Suzuki, M., Miura, S., Suematsu ,M. et al. (1992) Helicobacter pylori-associated ammonia production enhances neutrophil-dependent gastric mucosal cell injury. Am. J. Physiol. 263: G719-G725.

12. Yoshida, N., Granger, D.N., Evans Jr, D.J. et al.(1993) Mechanisms involved in Helicobacter pylori-induced inflammation. Gastroenterology 105: 1431–1440.

13. Furchgott, R.F. and Zawadzki, J.V.(1980) The obligatory role of endothelial cells in the relaxation of arterial smooth muscle by acetylcholine. Nature 288: 373–376.

14. Gryglewski, R.J., Palmer, R.M.J. and Moncada, S.(1986) Superoxide anion is involved in the breakdown of endothelium-derived vascular relaxing factor. Nature 320: 454–456.

15. Yoshikawa, T., Yasuda, M., Ueda ,S. et al.(1991) Vitamin E in gastric mucosal injury induced by ischemia-reperfusion. Am. J. Clin. Nutr. 53: 210S-214S.

16. Yoshikawa, T., Naito, Y., Tanigawa, T. et al.(1991) The antioxidant properties of a novel zinc-carnosine chelate compound, N-(3-aminopropionyl)-L-histidinato zinc. Biochimica et Biophysica Acta 1115: 15–22.

17. Yoshikawa, T., Takahashi, S., Ichikawa, H. et al.(1991) Effects of TJ-35(Sugyaku-san) on gastric mucosal injury induced by ischemia-reperfusion and its oxygen-derived free radical-scavenging activities. J. Clin. Biochem. Nutr. 10: 189–196.

3

CAROTENOIDS AS ANTIOXIDANTS

Junji Terao,[1] Syunji Oshima,[2] Fumihiro Ojima,[2] Boey Peng Lim,[3] and
Akihiko Nagao[1]

[1]National Food Research Institute,
Ministry of Agriculture, Forestry and Fisheries
Tsukuba 305, Japan
[2]Kagome Research Institute, Kagome Co.
Nishinasuno, Tochigi 329–27, Japan
[3]School of Chemical Sciences, Universiti Sains Malaysia
Penang 11800, Malaysia

1. INTRODUCTION

A wide variety of antioxidants are found in foods and foodstuffs. Some of them such as vitamin C and vitamin E are recognized to play an essential role in the antioxidant defense on the oxidative damages *in vivo*. Carotenoids seems to be a member of dietary antioxidants acting in human body. Epidemiological studies have strongly suggested that consumption of carotenoid-rich foods are inversely related to the incidence of cancer and other degenerative diseases. An intervention study in China[1] demonstrated that the supplementation of β-carotene, a typical carotenoid widespread in plant foods, is helpful in the prevention of cancer death. However, the role of dietary carotenoids in the antioxidant defense *in vivo* is still uncertain and therefore it is controversial whether or not antioxidant activity of carotenoids result in cancer-prevention.

Previous studies clarified that β-carotene acts as chain-breaking antioxidant in solution and membrane lipids by scavenging chain-propagating free radicals[2–4]. On the other hand, carotenoids are well known singlet oxygen quenchers[5]. Singlet oxygen-quenching activity of carotenoids may also participate in the antioxidant defense *in vivo* because this highly reactive oxygen species can oxidize biological components including membrane lipids. Furthermore, carotenoids consist of more than 500 species including hydrocarbon carotenoids such as β-carotene and lycopene and oxygenated carotenoids such as lutein and astaxanthin. Human absorbs these carotenoids unselectively and accumulates a variety of carotenoids[6]. Several questions are raised in the antioxidant activity of carotenoids *in vivo*. One of them is whether carotenoids are singlet oxygen quenchers or free radical scavengers *in vivo*. Second one is whether β-carotene or carotenoinds other than β-carotene are more important as dietary antioxidants.

Food and Free Radicals, edited by Hiramatsu *et al.*
Plenum Press, New York, 1997

To resolve these questions, we measured the effectiveness of β-carotene as free radical scavenger and singlet oxygen quencher in a biomembrane model. Then, singlet oxygen-quenching activity of β-carotene was compared to that of astaxanthin, a typical polar carotenoids present in Crustacea, Tunicates and fish. The effect of endogenenous carotenoids on the resistance of human plasma lipoprotein against oxidative attack was also measured by preparing carotenoid-rich plasma through the continuous supplementation of tomato juice to human volunteers. The results strongly suggest that dietary carotenoids, not only β-carotene, mainly act as singlet oxygen quenchers in the antioxidant defense *in vivo*.

2. MATERIALS AND METHODS

2.1. Materials

β-Carotene was obtained from Sigma Chemicals (St. Lous, MO., USA). Astaxanthin was kindly provided by Hoffman-La Roche (Basle, Swtzerland). *d*-α-Tocopherol was the gift from Eisai Co. (Tokyo, Japan). Egg yolk phosphatidylcholine (PC) was the product of Sigma Chemicals and purified by reverse-phase column chromatography. 2,2'-Azobis(2,4-dimethyl-valeronitrile)(AMVN) and 2,2'-azobis(2-amidinopropane)hydrochloride (AAPH) were obtained from Wako Pure Chemcal Industries (Osaka, Japan). Methylene blue was purchased from Kanto Chemical Co. (Tokyo, Japan) and 12-(1-pyrene)dodecanoic acid (p-12) was the product of Lambda Probes & Diagnostic (Graz, Austria).

2.2. Preparation of Carotenoid-Containing Liposomes

The chloroform solutions of PC and carotenoids were mixed in a test tube and solvent was removed with a stream of nitrogen and finally under vaccum. The residue was dispersed in Tris-HCl buffer (0.01M, pH 7.4) containing 0.5 mM diethylenetriamine pentaacetic acid (DTPA). Multilamellar liposomes (MLV) were obtained by mixing the suspension with a Vortex mixer followed by ultrasonic irradiation for 30 sec. Large unilamellar liposomes (LUV) were obtained from MLV by extrusion method as described below. The suspension was introduced to LiposoFAST (AVESTIN Co., Ottawa, Canada) and passed through polycarbonate membrane (pore size; 100 mm) 21 times. The resulting liposomal suspension was diluted with Tris-HCl buffer.

2.3. Free Radical Oxidation of Liposomal Membranes

A hexane solution of AMVN was mixed with PC before preparation of liposomes. MLV were prepared and the content was incubated in the dark at 37°C with continuous shaking.

2.4. Photosensitized Oxidation of Liposomal Membranes

In methylene blue sensitized photooxidation, the suspension of MLV was mixed with Tris-HCl buffer containing methylene blue. In P-12 sensitized photooxidation, a chloroform solution of P-12 was added to the chloroform solution containing PC and carotenoids before preparation of LUV. The suspension of LUV was placed into a chamber and photoirradiated by fluorescent lamp (light intensity at the sample: 10,000 lx) at 37°C with continuous shaking.

2.5. Determination of Carotenoids and α-Tocopherol

Carotenoids were quantified by reverse-phase HPLC using a column of TSK gel ODS 80T (TOSOH, Japan) with the eluting solvent of isopropanol/acetonitrile/dichloromethane (7:7:2, v/v/v) for β-carotene and methanol/water (95:5, v/v) for astaxanthin. The effluent was monitored by the absorbance at 450 nm. Reverse-phase HPLC with fluorophotometric detection or amperometric detection was used for the determination of α-tocopherol[7].

2.6. Determination of PC Hydroperoxides (PC-OOH)

PC-OOH were determined by a reverse-phase HPLC according to the method described previously[8,9].

2.7. Preparation of Carotenoid-Rich Low-Density Lipoprotein (LDL)

One volunteer (26 years-old, male) whose plasma was poor in carotenoid content was selected from 76 healthy volunteers working at the same place. The volunteer supplemented 160 g of tomato juice at his daily diet for 19 days. The amount of tomato juice per day was equivalent to single lycopene dose of 0.44 μmol/kg weight. At the end of the supplementation period, blood was collected and its plasma was immediately prepared by centrifugation. The LDL fractions were prepared by ultracentrifugation method from the plasma before and after supplementation[10].

2.8. Oxidation of LDL Suspension

The LDL suspension was diluted by PBS to adjust the final concentration of cholesteryl ester (1.57 μmol/ml) and mixed with methylene blue solution (final concentration of methylene blue: 1.0 mM). The mixed suspension was then photoirradiated by fluorescent lamp (light intensity at the sample; 10,000 lx) at 37°C with continuous shaking. In the case of radical generator-mediated oxidation of LDL, LDL suspension was mixed with the aqueous solution of AAPH (final concentration; 1.0 mM). The mixture was incubated in the dark at 37°C.

2.9. Determination of Cholesteryl Ester Hydroperoxides (CE-OOH) in LDL

An aliquot of the suspension (100 μl) was withdrawn and ethanol solution of *trans*-beta-8'-apocarotenal was added to the mixture as an internal standard. The solution of n-hexane and dichloromethane (4:1, v/v, 1.0 ml) was added to the mixture and centrifuged. The supernatant was evaporated and subjected to HPLC analysis. HPLC was run with UV detection at 235 nm as described previously[10]. The concentration of CE-OOH was tentatively calculated from the standard curve of the hydroperoxy derivative of cholesteryl linoleate.

3. RESULTS AND DISCUSSION

3.1. Activity of β-Carotene as Singlet Oxygen Quencher and Free Radical Scavenger within Liposomal Membranes

Fig 1 shows the accumulation of PC-OOH in lipid-soluble AMVN-induced oxidation of PC liposomes under air and low partial pressure of oxygen. The oxidation of PC was initiated by the peroxyl radical generated from AMVN within bilayer membranes and PC-OOH accumulated at a known and reproducible rate by the chain reaction mediated by PC peroxyl radicals[11]. Conventional antioxidants such as α-tocopherol can trap chain-propagating peroxyl radicals resulting in the retardation of rapid accumulation of PC-OOH[11].

Figure 1 shows that α-tocopherol effectively suppressed the PC-OOH accumulation either under air or low partial pressure of oxygen. β-Carotene also retarded the accumulatioin of PC-OOH under air and its retardation was enhanced under low partial oxygen pressure. However, its effectiveness was lower than that of α-tocopherol independent of the partial oxygen pressure.

Fig 2 shows the consumption of β-carotene and α-tocopherol during the AMVN-induced oxidation of liposomal lipids. Interestingly, consumption of β-carotene was suppressed by the coexistence of α-tocopherol, although consumption of α-tocopherol was little affected by the coexsitence of β-carotene. It implies that α-tocopherol suppresses the oxidative loss of β-carotene during the chain reaction of lipid peroxidation occurring in biomembranes. α-Tocopherol is much more effective than β-carotene in the scavenging of

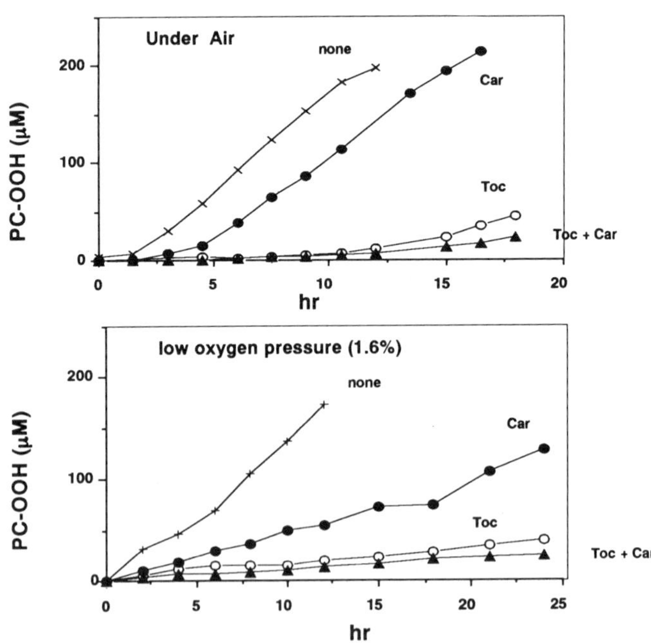

Figure 1. Effect of β-carotene and α-tocopherol on free radical-mediated oxidation of PC liposomes under air and low oxygen pressure. Car:β-carotene, Toc:α-tocopherol.

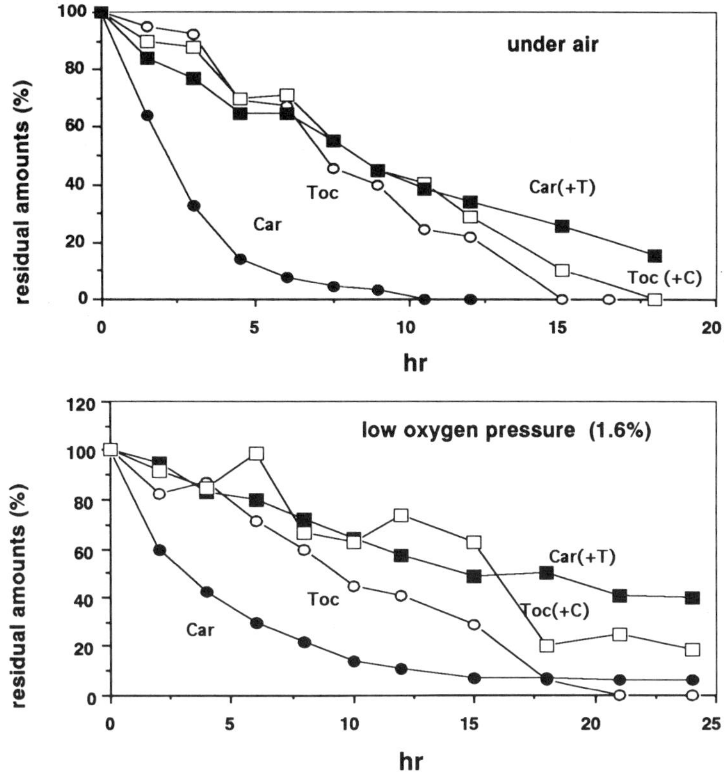

Figure 2. Decrease of β-carotene and α-tocopherol during the free radical mediated oxidation of PC liposomes. Car:β-carotene, Toc:α-tocopherol, Car(+T): β-carotene in the coexistence of α-tocopherol, Toc(+C): α-tocopherol in the coexistence of β-carotene.

of chain-propagating lipid peroxyl raidcals and thereby intercepts the radical attack to β-carotene.

Next, the effect of β-carotene on the singlet-oxygen mediated oxidation of liposomal phospholipids was compared with that of α-tocopherol using p-12-sensitized photooxidation (Type II) (Figure 3). This sensitizer generates singlet oxygen within the membraneous phospholipids because of its lipophilic character. In this system, β-carotene inhibited the accumulation of PC-OOH, whereas α-tocopherol scarcely affected its accumulation. Therefore, it is apparent that carotenoids effectively inhibit singlet oxygen-mediated lipid peroxidation when this reactive oxygen species is generated within the biomenbrane lipids. This is rationalized by the kinetic rate constant of singlet oxygen-quenching by carotenoids. The rate constant by β-carotene is 100 times higher than that of α-tocopherol[12]. Thus, carotenoids can act as excellent singlet oxygen quencher at lower concentration. We previously showed that β-carotene suppressed the consumption of α-tocopherol when lower amounts of β-carotene coexisted with higher amount of α-tocopherol in liposomal membranes[10]. β-Carotene may prevent the consumption of α-tocopherol by intercepting the attack of singlet oxygen to α-tocopherol in biomembranes.

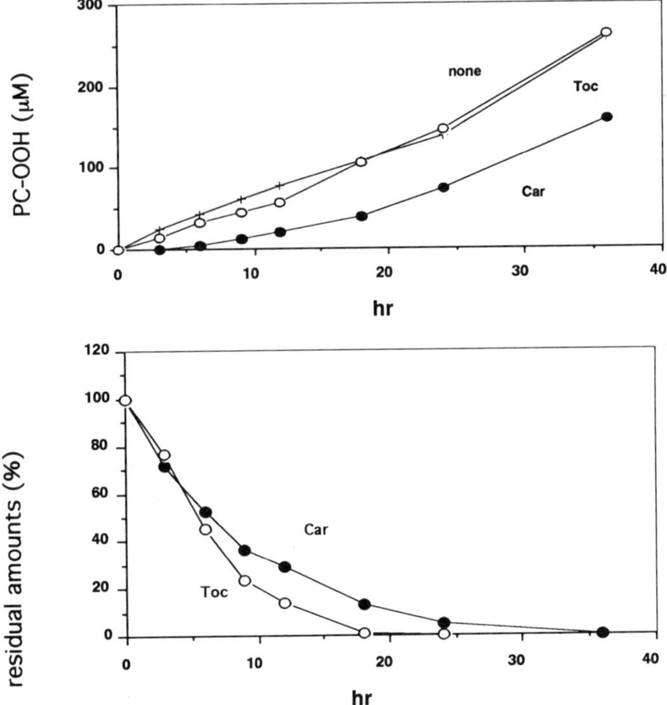

Figure 3. Effect of β-carotene and α-tocopherol on singlet oxygen oxygenation of PC liposomes.

3.2. Comparison of the Singlet Oxygen Quenching Activity between β-Carotene and Astaxanthin in Liposomal Membranes

We compared the effectiveness of β-carotene and astaxanthin as singlet oxygen quencher in liposomal membranes to understand the difference of the activity between nonpolar carotenoids and polar carotenoids (Fig. 4).

P-12 and methylene blue were used as the photosensitizers producing singlet oxygen. In the case of methylene blue-sensitized photooxidation, singlet oxygen was generated in the water-phase because of its water-soluble property. In both photosensitized

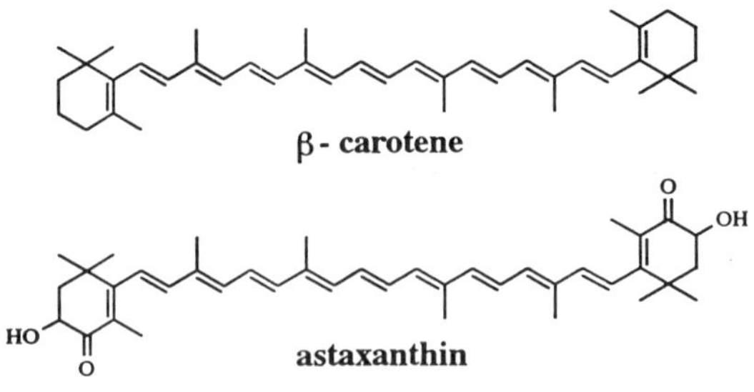

Figure 4. Structures of β-carotene and astaxanthin.

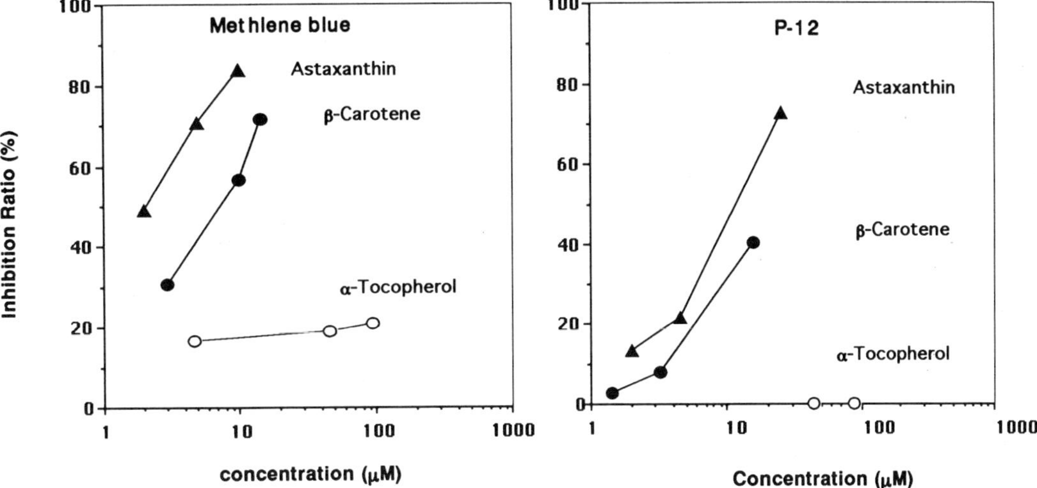

Figure 5. Concentration-dependent inhibition of singlet oxygen-mediated PC liposomes by carotenoids and α-tocopherol.

oxidation systems, inhibition of PC-OOH formation by astaxanthin was more effective than β-carotene at the lower concentration (Fig. 5).

This indicates that astaxanthin is more effective singlet oxygen quencher in the membrane system. Dimascio et al[13] demonstrated that quenching rate constant of astaxanthin was higher than that of β-carotene. Furthermore, we already clarified that astaxanthin was more resistant to photoirradiation itself and to chemical reaction with singlet oxygen[7]. This resistance may lead to effective inhibition of singlet oxygen oxidation in membrane phospholipids.

3.3. Resistance of Human Plasma LDL-Rich in Carotenoids against Oxidative Attack

The concentration of plasma carotenoids for the volunteer was increased from 0. 43 μM to 1.88 μM by the 19 day-continuous supplementation of tomato juice. Total carotenoid contents in LDL preparation also increased from 0.84 nmol/mg protein to 2.23 nmol/mg protein after the supplementation. In particular, the contents of lycopene and β-carotene were increased about 6.0-fold among the carotenoids present in LDL (Table 1).

Singlet oxygen-mediated oxidation of LDL was performed by mixing methylene blue solution with the LDL suspension. The amount of CE-OOH after photoirradiation in carotenoid-rich LDL suspension was lower than that in the control LDL indicating that carotenoi-rich LDL was more resistant to photosensitized oxidation than the control LDL. However, there was no significant difference in the amounts of CE-OOH between the two suspensions when they were exposed to AAPH-induced oxidation. This indicates that the oxidizability against AAPH-derived radicals is indistingushable between carotenoid-rich LDL and control LDL.

Wagner et al.[14] have shown that lycopene and β-carotene are included in the principal singlet oxygen quenchers within human LDL and suggested that beneficial effect of these compounds in health may in part be due to the elimination of singlet oxygen occur-

Table 1. Oxidation of carotenoid-rich LDL suspension and control LDL suspension

	Control LDL	Carotenoid-rich LDL
Carotenoids (ng/ml)	229	387
Lycopene	28	125
β-Carotene	38	150
α-Tocopherol (μg/ml)	5.19	4.91
Choltesteryl ester (mg/ml)	1.01	1.01
CE-OOH accumulated by singlet oxygen oxidation[a]	58.9 ± 4.5*	41.7 ± 4.6*
CE-OOH accumulated by radical oxidation[a]	38.2 ± 0.6	40.2 ± 3.4

*significant difference (p<0.01)
[a]LDL suspension was incubated for 4 hrs.

ring *in vivo*. The present study has clearly shown that carotenoid-rich LDL is more resistant to singlet oxygen-mediated oxidation than carotenoid-poor LDL. It has been demonstrated that phagocytosing neutrophils[15] and stimulated macrophases can generate this oxygen species[16]. It seems plausible that LDL in the blood stream or within the arterial wall tend to be in the exposure of singlet oxygen. The present study first demonstrated the supplementation of LDL with carotenoids *in vivo* increases the resistance against the singlet oxygen oxidaiton of LDL. On the other hand, we could not find the inhibition by the supplementation of LDL with carotenoids on the free radical attack of LDL. It is unlikely that the supplementation of carotenoids increases the free radical scavenging activity of LDL although further study is required to know the effect of carotenoids in free radical attack to LDL.

4. CONCLUSION

It is likely that carotenoids distributed in membraneous phospholipids and plasma lipoproteins mainly act as singlet oxygen quenchers. Carotenoids other than β-carotene also participate in the antioxidant defense *in vivo* and therefore their antioxidant activity should be clarified.

REFERENCES

1. Blot, W., J., Li J-Y., et al. (1993) Nutrition InterventionTrail in Linxian, China: Supplementation with specific vitamin/mineral combinations. J. Natl. Cancer Inst. 85:1483.
2. Burton G.W. and Ingold, K. U. (1984) β-Carotene:An unusual type of lipid antioxidant. Science 224: 569–573.
3. Kennedy, T.A., and Liebler D.C. (1992) Peroxyl radical scavenging by β-carotene in lipid bilayers J. Biol. Chem.267:4658–4663.
4. Pallozza, P., Moualla, S., Krinsky, N.I. (1992). Effect of β-carotene and α-tocopherol in radical initiated peroxidation of microsomes. Free Radical Biol. Med. 13:127–136.
5. Foote C.S., and Denny R.W. (1968) Chemistry of singlet oxygen VII. Quenching by β-carotene J. Am. Chem. Soc. 90:6233–6235.
6. Goodwin, T. W. (1984) Mammals. In: The Biochemistry of the Carotenoids Vol. II pp.193–195, Chapman and Hall, London.
7. Oshima, S., Ojima, F, Sakamoto, H., Ishiguro, Y. and Terao, J. (1993) Inhibitory effect of β-carotene and astaxanthin on photosensitized oxidation of phospholipid bilayers. J. Nutr. Sci. Vitaminol. 39: 607–615

8. Terao, J., Nagao, A., Park, D.K., and Boey, P.L. (1992) Lipid hydroperoxide assay for antioxidant activity of carotenoids. Methods Enzymol. 213:454–460.

9. Terao, J., Asano, I., and Matsushita S. (1985) Preparation of hydroperoxy and hydroxy derivatives of rat liver phosphatidylcholine and phosphatidylethanolamine. Lipids 20:312- 317.

10. Ojima, F., Sakamoto, H., Ishiguro, Y., and Terao J. (1993) Consumption of carotenoids in photosensitized oxidaiton of human plasma and plasma low-density lipoprotein. Fee Radical Biol. Med. 15: 377–384.

11. Niki, E. (1990) Free Radical Initiators as source of water and lipid- soluble peroxyl radicxals. Methods in Enzymol. 186:100–108.

12. Terao, J., Yamauchi, R., Murakami, H., and Matsushita, S. (1980) Inhibitory effect of tocopherols and β-carotene on singlet oxygen- initiated photooxidation of methyl linoleate and soybean oil J. Food Process Preserv. 4:79–93.

13. Dimascio, P., Kaiser, S., and Sies H. (1989) Lycopene as the most efficient biological carotenoid singlet oxygen quencher. Arch. Biochem. Biophys. 280, 1–8.

14. Wagner, J. R., Motchnik, P. A., Stocker R., Sies, H. and Ames, B. N. (1993) The oxidation of blood plasma and low density lipoprotein components by chemically generated singlet oxygen J. Biol. Chem. 268:18502–18506.

15. Steinbeck, M.J., Khan, A.U., and Karnovsky, M.J. (1993) Extracellular production of singlet oxygen by stimulated macrophages quantified using 9,10-diphenylanthracene and perylene in a polystyrene film. J. Biol. Chem. 268:15649–15654.

16. Steinbeck, M.J., Khan, A.U., and Karnovsky, M.J. (1992) Intracellular singlet oxygen generation by phagocytosing neutrophyls in response to particles coated with a chemical trap. J. Biol. Chem. 267: 13425–13438.

4

PHENOLIC ANTIOXIDANTS

Takuo Okuda

Faculty of Pharmaceutical Sciences
Okayama University
Tsushima, Okayama 770, Japan

1. INTRODUCTION

Various polyphenolic compounds isolated from the plants used as medicine and food have antioxidant activities[1–3], and there are marked differences among their activities depending on their chemical structures and the experimental systems[4,5].

2. EXAMPLES OF THE PLANTS USED AS BOTH MEDICINE AND FOOD

Besides the medicinal plants which are unsuitable for food because of the presence of the components of drastic activity such as some alkaloids which make the plant a poisonous one, there are also a variety of plants which have been used as both medicine and food. It is notable that there are a number of polyphenol-rich plants among the latter type of plants. Some examples of the polyphenol-rich plants and their products used as food and/or medicine in the world are in Table 1.

3. IS THERE DIFFERENCE AMONG POLYPHENOL, TANNIN AND TANNIC ACID?

Polyphenols consist of a large number of the organic compounds of various chemical structures, having multiple phenolic hydroxyl groups in the molecule. Thus, a wide range of compounds, mostly natural products, are covered under the name polyphenol; hydrolyzable tannins (=metabolites of gallic acid), condensed tannins (=proanthocyanidins), flavonoids, coumarins, lignans, lignins, quinoids, and phenolic compounds of small molecule such as pyrogallol, catechol and catechin, etc.

Recently, the name "plant polyphenols" is sometimes used in place of tannin which is the largest group of compounds among natural polyphenolic compounds. There may be

Food and Free Radicals, edited by Hiramatsu *et al.*
Plenum Press, New York, 1997

Table 1. Examples of polyphenol-rich plants used as food and medicine

Tea (green tea, black tea, etc.), red wine, grape, Kaki, cinnamon, cloves, chestnut, peach, walnut, rose, strawberry, eucalyptus, guava, myrobalans, *Agrimonia* sp, *Saxifraga stolonifera, Geranium thunbergii*	(proanthocyanidins and hydrolyzable tannins)
Coffee, artichoke, dandelion, *Artemisia* sp., *Chrysanthemum* sp.	(caffetannins)
Basil, hyssop, lavender, lemon balm, marjoram, mint, pomegranate, raspberry, rosemary, sage, thyme, Shiso (*Perilla* sp.)	(labiataetannins)
Licorice and most of vegetables and fruits	(flavonoids and other polyphenols)

some reasons for this substitution of the name. The name tannin has long been used for intractable mixtures of phenolic materials often found in plant extracts, without meaning any compound of defined chemical structure. It happened sometimes that an unfavorable biological activity found for a plant was attributed to "tannin", without knowing what is the tannin in that plant, probably because it is an easy way of answering question, to attribute the activity in question to an unknown material in the plant extract. This situation has recently been changed remarkably by the progress in the chemistry of tannin. An example is the investigation of the effects of (-)-epicatechin gallate (EGCG), which is the main component of "green tea tannin"[34].

Today, "tannin" means each compound of clearly determined chemical structure belonging to the tannin family, as well as the whole family of tannin. This has been caused by the isolation of hundreds polyphenols (the number of isolated tannins may now be over a thousand depending on the counting way) from various plants, followed by the elucidation of their chemical structures. The discrimination of tannins in various plant species, and their chemical classification have thus been established to a large extent.

Therefore the name "plant polyphenols", as a synonym of "tannin", can be used now in a way similar to that of the other natural organic compounds, such as those belonging to alkaloids, terpenoids, steroids, etc. Today, any argument concerning the biological activities of tannin or polyphenol should be based on each compound of defined chemical structure in this class.

As for the name "tannic acid", practically there was no difference between this name and "tannin", in the old time when their chemical evidence was not available. After some partially purified tannin preparations named "tannic acid", which are prepared from Chinese gall, Turkish gall, etc., appeared in the market, the name "tannic acid" meant these preparations, besides the tannin in general. The "tannic acid" of the former definition, which was pracitically the only tannin preparation available for scientists, were in fact the mixtures of many polyphenols, mostly of gallotannins. They exhibit many peaks of their components in the HPLC chart of normal and reversed phase developments. However, these "tannic acid" preparations are examples of tannins of rather better purity and identification than most of the "tannins" referred in the old literatures.

4. ACTIVITIES OF THE POLYPHENOLS OF DEFINED CHEMICAL STRUCTURES

The research of biological activities of tannins or polyphenols in recent years[1-3] are those carried out with each pure polyphenol of determined chemical structure. The correlation between each polyphenol and the plant species producing it in the plant taxonomy has also been widely established[1–3,9,12].

5. THE DIFFERENCE BETWEEN TANNIN AND OTHER POLYPHENOL - BINDING WITH PROTEIN

The major known difference between tannin and polyphenol of the other types such as flavonoid is as follows. Tannins are polyphenols which bind with protein, basic compounds such as alkaloids, or heavy metalic ion in a solution, making them insoluble and inducing precipitation. The activity binding with protein is considered to be mainly due to formation of multi-hydrogen bonds.

The precipitate formation with protein is a characteristic property of regular tannins which are the polyphenols of fairly large molecule, but there are also some polyphenols which show potent binding activity in spite of their small molecules. An example is EGCG, the main polyphenol of green tea, which shows potent anti-tumor-promoting activity[34] besides strong binding activity to protein[6]. Several flavonoidal compounds isolated from licorice also exhibit the binding activities, although lower than that of EGCG[7].

6. CLASSIFICATION OF POLYPHENOLS CALLED TANNIN

The polyphenols called tannin or related to tannin, can be classified as in Table 2. Some of the polyphenols found in the plants used as food or medicine are also in this table.

The tannin content in some plants is over 10% , and is over 50% in some specific plant parts such as Chinese gall and Turkish gall. We often use tannin-rich plants in everyday life, as food, beverage, or health drinks, and these tannin-rich plants have been taken in by numerous people for long years without finding any notable toxicity.

Table 2. Classification of polyphenols called "tannin"

Type of tannin	Example of tannin
A. Hydrolyzable tannins	
A-1. Gallotannins	Pentagalloylglucose
	Polygalloylglucoses in galls
A-2. Ellagitannins and metabolites	Geraniin, Corilagin, Pedunculagin
A-3. Ellagitannin oligomers	Agrimoniin, Oenothein B
B. Condensed tannins	
B-1. Proanthocyanidins	
Procyanidins	Epicatechin oligomers
Prodelphinidins	Epigallocatechin olygomers
Others	
B-2. Galloylated proanthocyanidins	Tannins in Kaki and *Saxifraga stolonifera*
C. Polyphenols of small molecule	EGCG, ellagic acid, gallic acid
D. Caffeates	
D-1. Caffetannins	Di-*O*-Caffeoylquinic acid
D-2. Labiataetannins	Rosmarinic acid, Rabdosiin

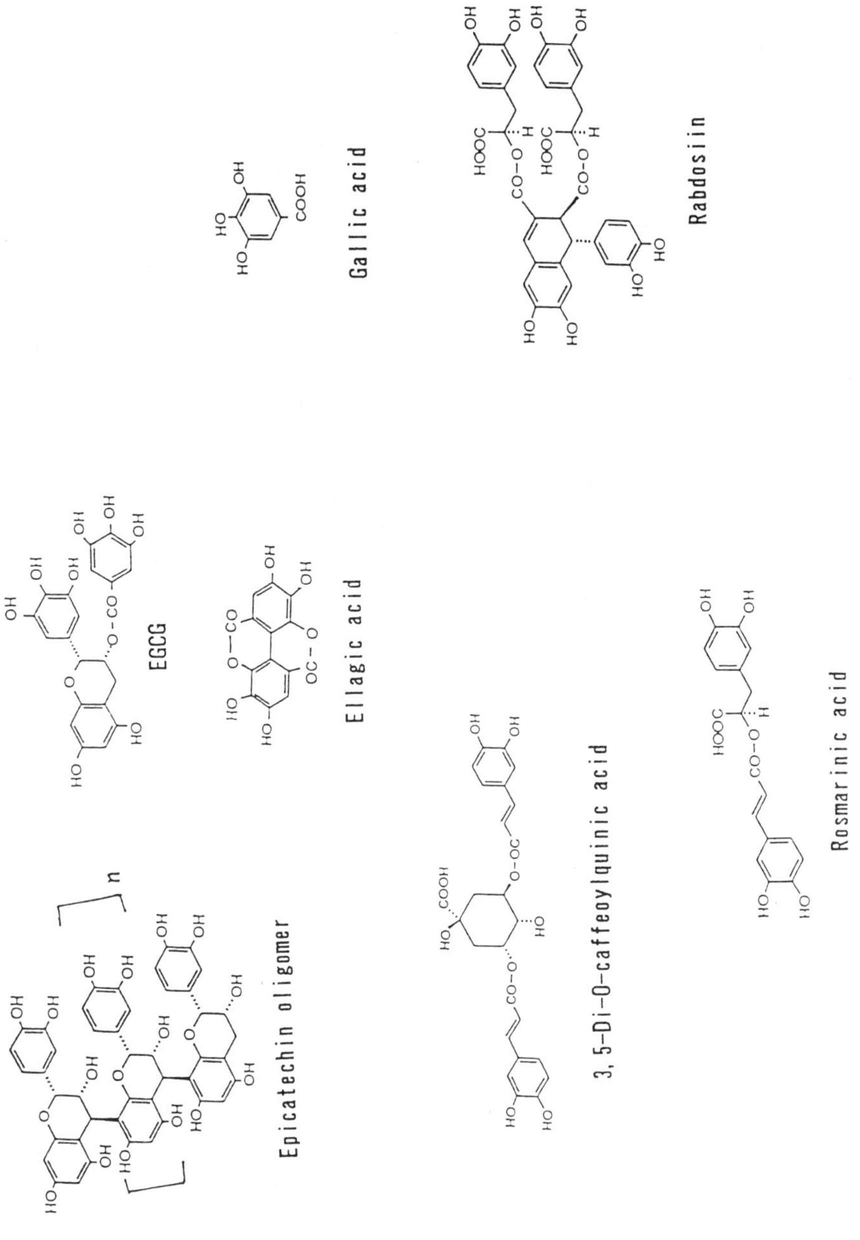

Gallic acid

Rabdosiin

EGCG

Ellagic acid

Epicatechin oligomer

3,5-Di-O-caffeoylquinic acid

Rosmarinic acid

Chart 1. Chemical structures of various tannins.

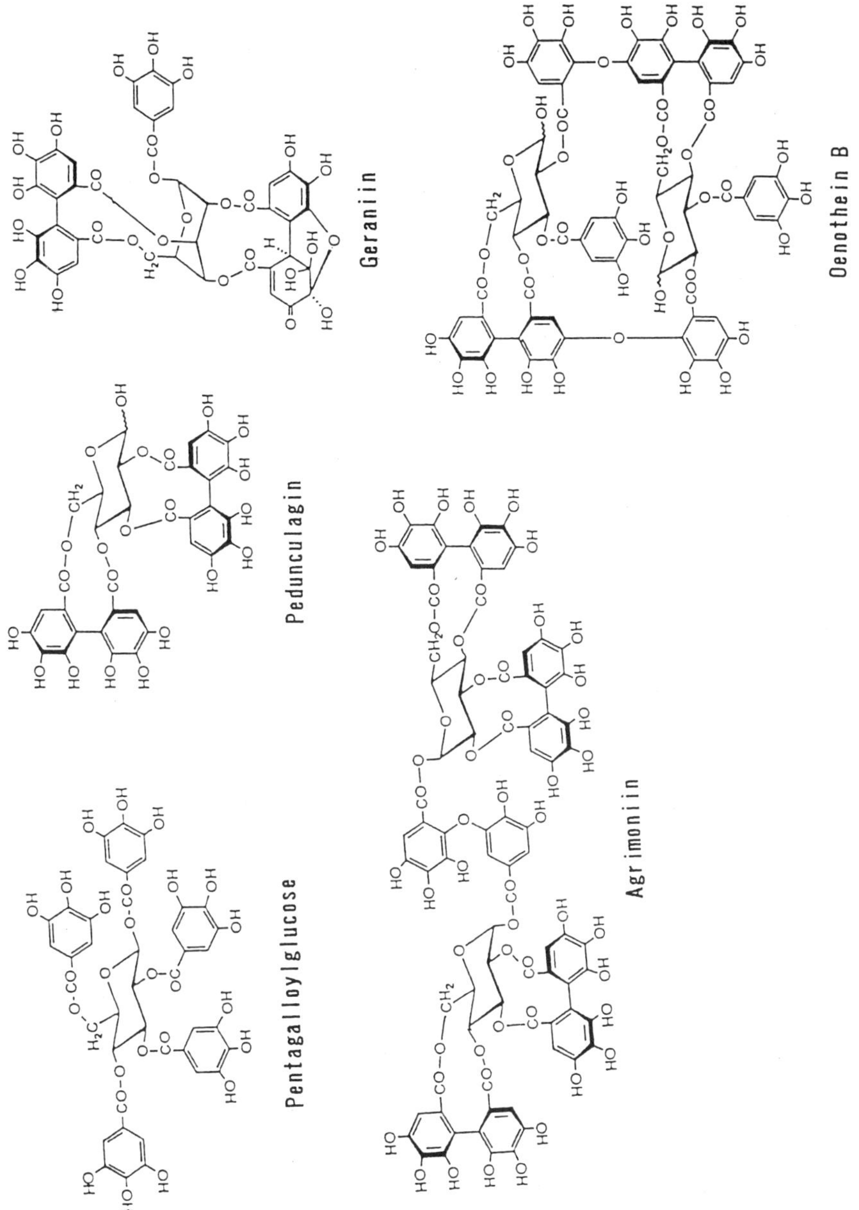

Chart 1. (*Continued.*)

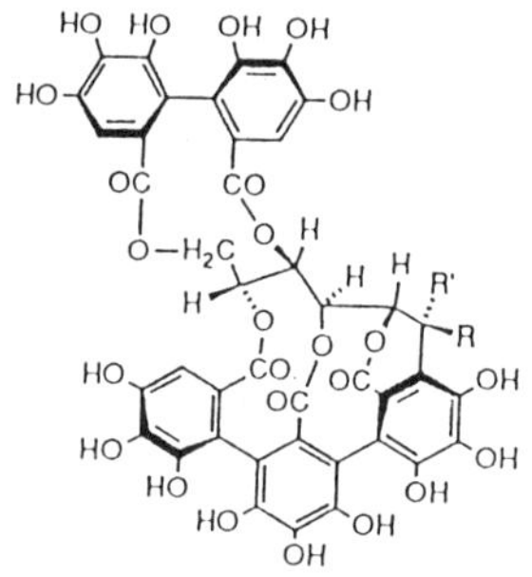

Castalagin: R=OH, R'=H
Vescalagin: R=H, R'=OH

Chart 2. Chemical structures of castalagin and vescalagin.

7. EXAMPLES OF TANNIN-RICH PLANTS USED AS FOOD AND MEDICINE

7.1. Grape, Wine, and Brandy

Wine, particularly red wine, is rich in tannin mainly coming from grape skin. The tannin in wine is removed when the wine is distilled to produce brandy, but storage of brandy in the barrel made of oak wood induces dissolution of ellagitannins represented by castalagin and vescalagin, which were contained in the barrel wood, into brandy[8]. During aging of the brandy, degradation of these polyphenols occurs, and ellagic acid is one of the final products from these ellagitannins. Lignins which are polyphenolic polymers in the barrel wood dissolve more slowly during the aging, and are the main polyphenols in the aged brandy[8]. The wood of *Quercus robur,* generally called oak, is the main source of preparing the barrel in Europe.

7.2. Cinnamon and Kaki

There are some fruits and spices which are rich in tannins. "Kaki", a fruit popular and indigenous to East Asia, and is nearly related to persimmon, is an example. It abundantly contains partially galloylated condensed tannins, even when it is ripe and sweet. Kaki's mature fruit is sweet and edible in spite of the presence of a large amount of partially galloylated condensed tannins. This phenomenon is due to the formation of large polymers of condensed tannin upon ripening of the fruit, which does not allow the tannin

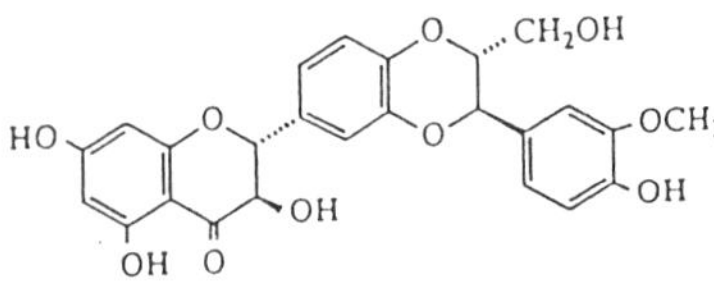

Silybin

Chart 3. Chemical structure of silybin.

Table 3. Caffetannin contents in *Artemisia* and coffee

Plant Sources	3,5-DCQ	3,4-DCQ	4,5-DCQ	Chlorogenic Acid (%)
(*Artemisia* species)				
A. princeps	18	3.5	1.8	6.4
A. montana	12	3.3	1.4	4.6
A. capillaris	5.4	0.88	1.2	4.5
(Coffee beans)				
Guatemala, dried	1.4	0.94	0.54	17
Guatemala, roasted	0.22	0.57	0.51	11
Tanzania, dried	1.5	1.1	0.45	25
Tanzania, roasted	0.25	0.25	0.25	14

to come out of the cells to give the astringent taste like the tannin in immature Kaki fruit having tannins of smaller molecules.

Cinnamon, which is frequently used as a spice and an ingredient in the traditional Kampo medicine prescriptions, is also one of the official medicines in Japan. This is also an example of condensed-tannin rich plant, although cinnamaldehyde, the main constituent of the essential oil in this plant, has usually been attracting more attention by its flavor.

7.3. Medicinal Plants of *Geranium* Species

Condensed tannins and hydrolyzable tannins are two large subfamilies in the tannin family, as shown in Table 2. The structural features distinguishing them from each other are as follows.

Condensed tannins are mutually condensed flavan derivatives which belong to flavonoids. The oligomer of (-)-epicatechin in Chart 1 is an example. They are often called proanthocyanidins because they give a pink-red color due to formation of anthocyanidins, when heated in acidic solution (*n*-butanol containing hydrochloric acid is often used). The flavan structural units in the condensed tannin molecule are bound with each other by a carbon-carbon bond which can be produced in the presence of acid or some enzyme, and the molecule of condensed tannin can be enlarged by further condensation in the presence of these catalysts. Isolation of each proanthocyanidin of various degree of condensation from each species of plant is a laborious work, and is sometimes impossible because of their molecular similarity to each other, and their unstableness. Therefore it may be adequate to say that condensed tannins still have some aspects rather in accord with the old concept of tannin, although their structural specificities to various plant species are observable.

Hydrolyzable tannins, whose structural examples are in Charts 1 and 2, constitute another big family of tannins. Each molecule of hydrolyzable tannin monomer has a carbohydrate or polyalcohol core which are bound with several polyphenolic carboxylic acids *via* ester linkages. Hydrolyzable tannins are hydrolyzed in the presence of alkali rather easily, and also in the presence of acid or enzyme, and even in hot water. Against the old concept of tannin, each hydrolyzable tannin is usually obtained as a pure compound of defined chemical structure, and produced by specifc species of plants, in a way similar to most of the other types of natural organic compounds, such as alkaloids, terpenoids and steroids.

Table 4. Rosmarinic acid content in labiate plants

Plant species (Fresh leaf)	Rosmarinic Acid (%)
Agastache rugosa	0.90
Ajuga decumbens	0
Glechoma hederacea var. *grandis*	1.48
Mentha arvensis var. *piperascens*	0.71
M. spicata var. *crispa*	0.63
M. lotundifolia	0.36
M. gentilis	0.36
Perilla frutescens var. *crispa*	1.21
P. frutescens var. *crispa* f. *viridi-crispa*	1.03
Prunella vulgaris var. *lilacina* (bud/leaf)	1.93/1.74
Rabdosia japonica (Dry leaf)	2.03
Ocimum basilicum (Sweet basil)	1.21
Origanum majorana (Marjoram)	2.95
Origanum vulgare (Oregano)	2.21
Rosmarinus officinalis (Rosemary)	2.36
Salvia officinalis (Sage)	2.29
Thymus vulgaris (Thyme)	1.32

Ellagitannin, a subgroup of hydrolyzable tannin, is remarkable for their extensive structural varieties[1,9]. Geraniin, one of the key compounds among ellagitannins, structurally, biogenetically and also in medicinal applicability, was first isolated from *Geranium thunnbergii* which is one of the most frequently used (mainly for digestive organs) medicinal plants in Japan. Geraniin is the main component of this plant of the content higher than 10% in dried leaf of this species, and distributes in a wide range of plants. Isolated geraniin forms crystals of yellow color similar to that of many flavonoidal compounds[10,11].

The oligomeric hydrolyzable tannins of various structures up to tetramers have been isolated recently from many species of plants[12]. Agrimoniin is the first oligomer isolated from plant[13], and oenothein B is an example of a dimer having macrocyclic structure[14]. Both of them have host-mediated antitumor activity[14]. The molecular weight of nobotanin K isolated from *Heterocentron roseum*, an example of tetrameric ellagitannin, is 3745.

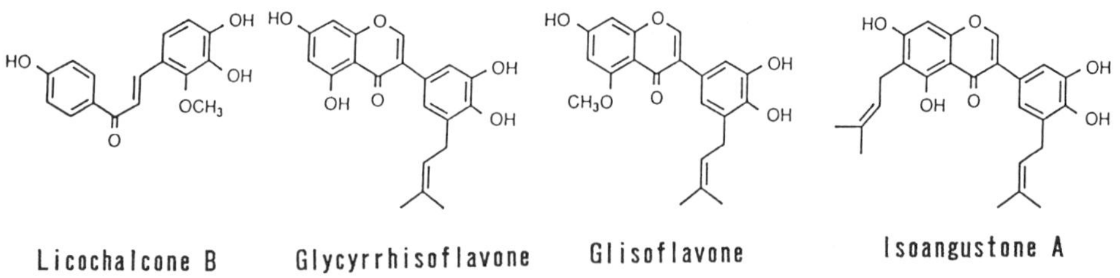

Chart 4. Examples of licorice phenolics structures.

7.4. *Artemisia* and Other Composite Plants

There are many species of plants in which the main components are polyphenol esters composed of caffeic acid bound to quinic acid or caffeic acid congener. They are called caffetannins and labiatetannins, depending on their occurrences or chemical structures. Antioxidant effects, one of the main effects of polyphenols in general, have also been found for caffetannins and labiataetannins.

Young leaf of Yomogi (*Artemisia princeps*) and Oyomogi (*A. montana*), belonging to Compositae, is eaten in Japan, often mixed in rice cake, and its mature leaf is used as a traditional Kampo medicine to stop bleeding and inflammations. The main components in these *Artemisia* species of plants are caffetannins represented by 3,5-di-*O*-caffeoylquic acid[15]. There is a great contrast between "*Artemisia*" and coffee concerning the contents of the components of caffetannins, as shown by the ratio of dicaffeoylquinic acids and of chlorogenic acid in Table 3. Although the name caffetannin came from the polyphenols contained in coffee, the main component in the caffetannin in coffee is chlorogenic acid whose binding activity is almost negligible. The binding activity of the total extract of coffee is thus merely 1/10 of the total extract of *Artemisa princeps* and *A. montana*. It may sound pradoxical, but as far as the binding activity which represents the properties of tannins is concerned, the caffetannin in coffee is not a tannin, while the di- and tri-caffeoylquinic acids in *Artemisia princeps* and *A. montana* are the main active components of these plants which have been regarded as tannin-rich plants, and may be called tannin, or caffetannin[15].

An example of the composite plants used as both food and medicine in Europe is artichoke (*Cynera scolymus*). This plant is a vegetable for salad, and is rich in polyphenols. The main component, 1,3-di-*O*-caffeoylquinic acid, which is structurally isomeric to 3,5-di-*O*-caffeoylquinic acid contained in Yomogi (*Artemisia princeps*), is a cholagogue and an antioxidant. Another example of composite plant used as a cholagogue in Europe is *Silybum marianum* (Mariendistel in German), containing silybin (the main component of silymarin obtainable by extracting this plant) which is a flavonolignan having cholagogic activity[16]. The extract of this plant has been used for treating disorders of liver, bile and stomach.

7.5. Labiate Plants

There are a variety of labiate plants used for cooking, cosmetic production, and everyday medical care. When classified according to the plant taxonomy, the largest family of the herbs traditionally used in Europe for food and medicine is Labiatae. There are also some Japanese labiate plants frequently used as both food and medicine, as exemplified by Shiso (*Perilla frutescens* var. *acuta*)[17]. Most of these labiate plants contain essential oil which is usually regarded as their characteristic component based on its fragrance, but the

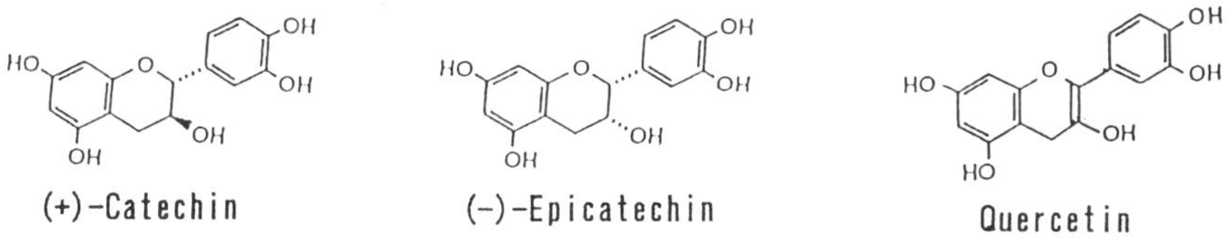

(+)-Catechin   (-)-Epicatechin   Quercetin

Chart 5. Chemical structures of catechin, epicatechin and quercetin.

Table 5. Effect on CCl_4-induced cytotoxicity in primary cultured rat hepatocytes

Polyphenols	GPT activity, %
Control	100
(Polyphenols of small molecule)	
Pyrogallol	30
Gallic acid	28
(Components of condensed tannins)	
(+)-Catechin	81
(-)-Epicatechin (EC)	80
(-)-Epigallocatechin (EGC)	34
(-)-Epicatechin gallate (ECG)	8
(-)-Epigallocatechin gallate (EGCG)	8
EC dimer	54
EC dimer monogallate	16
EC dimer digallate	12
(Hydrolyzable tannins)	
Pentagalloylglucose	6
Tellimagrandin I	2
Tellimagrandin II	5
Pedunculagin	5
Casuarictin	3
Geraniin	8
Dehydrogeraniin	25
Furosinin	28
Isotechebin	1
Rugosin A	2
Rugosin C	3
Agrimoniin	6
Gemin A	9
(Polyphenol of other type)	
Bergenin	104

components often found most abundant in these plants are polyphenols which are sometimes called labiataetannin[17].

The main polyphenol in many labiate plants is rosmarinic acid having a dimeric structure of caffeic acid[17,18]. The polyphenols of the structures corresponding to trimers and tetramers of caffeic acid have also been isolated from these plants. Melitric acids A and B in *Melissa officinalis* are examples of the trimers[19], and rabdosiin in *Rabdosia japonica* is an example of tetramers of caffeic acid[20].

7.6. Licorice

The root and rhizome of licorice is utilized as a sweetening agent, and is the most frequently used drug among the ingredients in the Kampo medicine prescriptions[21]. Glycyrrhizin is known as the main component of licorice, but licorice is also rich in phenolic compounds of various chemical structures. These phenolics are active as antioxidants, and exhibit inhibitory activities on XOD[22] and MAO[23], referred in Chapters 8 and 9. Some of these phenolic compounds also have fairly strong binding activity to protein.

Examples from a large number of the licorice phenolics, shown by chemical structures, are in Chart 4.

8. POLYPHENOLS AS ANTIOXIDANTS

Because most of flavonoids, which widely distribute in plants including those used as vegetables and fruits, are structurally polyphenols, the antioxidant activities of tannins, flavonoids and the other types of natural polyphenols can be discussed to some extent on the same chemical basis. Flavans having phenolic hydroxyl groups, exemplified by (+)-catechin and (-)-epicatechin and are the constructing units of condensed tannin molecules, belong to flavonoids, and the main structural difference of these flavans from most of the other types of flavonoids is the absence of carbonyl group in the flavan structures.

Flavonoids and tannins are similar to some other antioxidants such as ascorbic acid and α-tocopherol, concerning the antioxidant effects attributable to the radical scavenging activity[24,25.] The similarity is also found in their behavior to accelerate oxidation of co-existing substances in the presence of high concentration of ferrous ion or cupric ion.

Besides the number of phenolic hydroxyl groups on a benzene ring and the presence or absence of a conjugated double bond which makes an enol group, mutual spacial location of two or more of such polyhydroxylated benzene rings in the molecule of flavonoids and flavan derivatives often induce marked differences of the effects. The antioxidant activity of the flavan monomer, such as (+)-catechin and (-)-epicatechin which have a saturated ring lacking carbonyl group in the center of molecule, is poor. On the other hand some flavonol, for instance quercetin, having the same number of hydroxyl groups as these flavans and also an enol group conjugated with carbonyl group, exhibits potent activity. However, the activity of the oligomers of the flavans is intensified with the increase of the extent of oligomerization, and markedly reinforced by the presence of galloyl group at O-3 of the ring located at the center of each monomer molecule. Examples of catechin analogs having potent antioxidant activities are (-)-epigallocatechin gallate (EGCG) and (-)-epicatechin gallate (ECG), and those lacking potent antioxidant activity are (-)-epicatechin and (-)-epigallocatechin. The remakable difference between these two types of compounds is observable in their ESR spectra. The former two compounds which are gallates exhibit remarkably intense ESR signals than the latter types of the compounds lacking galloyl group, showing the significance of the presence of a galloyl besides a pyrogallol or catechol group. These phenomena are attributable to high stability of free radical generated from the galloyl group in the molecule of EGCG and ECG.

The presence of multiple galloyl group and its metabolites in the fairly large molecule of polyphenols, represented by those of hydrolyzable tannins, is also considered to be one of the structural features stabilizing their free radicals, which enable these polyphenols to be potent antioxidants.

The inhibition of lipid-peroxidation in mitochondria and microsomes of rat liver was one of the experimental evidence in which the antioxidant effect of plant polyphenols was observed. The polyphenols of various chemical structures isolated from medicinal plants inhibited lipid peroxidation induced in mitochondria by ADP plus ascorbic acid, to different extents[26]. Similar inhibitory effect of the polyphenols was also observed on lipid-peroxidation caused in microsomes by ADP and NADPH.

Sometimes a question may be raised concerning whether this kind of activity is due to the binding of the polyphenols to some material such as enzyme and protein in the experimental system, because tannins usually bind well with protein, enzyme, metalic ions,

Table 6. Effects of polyphenols on DPPH
radical

	IC_{50} after 30min.
Pentagalloylglucose	3.2
Tellimagrandin II	4.2
Pedunculagin	5.6
Geraniin	5.9
Chebulinic acid	5.3
Corilagin	6.8
EGCG	6.7
ECG	2.0
EGC	11.0
EC	31.0
(+)-Catechin	34.0
Gallic acid	16.0
Ascorbic acid	39.0
dl-α-Tocopherol	37.0

and various polymers. However, when relative extents of the binding activities to protein of these polyphenols were compared with those of their inhibitory effects on lipid peroxidation, no correlation was found between these two activities. Therefore, the antioxidant effects of these polyphenols should mainly be attributed to some other activity.

The participation of the radical-scavenging activity of the polyphenols in the inhibition of the lipid peroxidation was shown by the following model experiment[25].

In this model experiment of lipid peroxidation, methyl linoleate was used as the substrate, and the peroxidation was initiated by the AIBN radical generated by irradiating AIBN in the solution. The polyphenols added to the solution inhibited the radical-chain reaction leading to the peroxidation of methyl linoleate to various extent, proving their activities as the radical scavenger. The activities of the polyphenols of moderate or large molecule were mostly markedly longer-lasting than those of ascorbic acid and α-tocopherol, indicating the higher stability of the polyphenol radicals[25].

Again there were marked differences between the relative intensities of the inhibitory activities and those of the binding activities of each polyphenol, and this result of the comparison was in accord with the radical-scavenging mechanism suggested in this chapter. Among the polyphenols applied to this experiment, those having a biphenyl group (hexahydroxydiphenoyl group) generally showed stronger inhibitory activities than the polyphenols of the other types. The activties of the polyphenols of small molecule were generally low.

The data in Table 5 show the inhibition of CCl_4-induced cytotoxicity in the primary-cultured rat hepatocytes, which is attributable to the radical-scavenging activity of polyphenols[27]. Generally the polyphenols having a hexahydroxybiphenyl (HHDP) group

Chart 6. Coupling of alkyl gallate radicals generated in the presence of DPPH to yield dialkyl hexahydroxydiphenate.

Table 7. Inhibitory effects of polyphenols on xanthine oxidase

	Molecular weight	IC_{50} (μM)	Binding activity $RAG_{(Geraniin=1)}$
(Hydrolyzable tannins)			
Gallotannins			
1,2,6-Tri-*O*-galloylglucose	636.5	39	0.64
1,2,3,6-Tetra-*O*-galloylglucose	788.6	12	1.11
1,2,4,6-Tetra-*O*-galloylglucose	788.6	8.1	
1,2,3,4,6-Penta-*O*-galloylglucose	940.7	3.3	1.29
Ellagitannin monomers			
Corilagin	643.5	>40	0.17
Pedunculagin	784.6	18	0.24
Casuarictin	936.7	11	0.59
Tellimagrandin II	938.7	3.1	0.93
Rugosin A	1106.8	8.0	1.08
Isoterchebin	954.7	14	0.91
Ellagitannin dimers & trimers			
Cornusiin A	1571.1	27	1.19
Oenothein B	1569.1	20	
Rugosin D	1875.3	3.1	1.02
Coriariin A	1875.3	9.8	0.95
(Caffeates)			
3,5-Di-*O*-caffeoylquinic acid	516.5	34	0.20
Rabdosiin	718.6	>40	
(Small polyphenols and others)			
Gallic acid	170.1	24	0.11
Ellagic acid	302.2	3.1	0.14
Valoneic acid dilactone	470.3	0.8	0.74

in the molecule (ellagitannins) showed higher potency than the polyphenols of other types, although the activities of the compounds having further oxidized type of the HHDP group, such as furosinin, were lower.

The effects of polyphenols on the arachidonate metabolism in rat peritoneal polymorphonuclear leukocytes were also investigated. The formation of 5-HETE, one of the lipoxygenase products, was potently inhibited by polyphenols (geraniin, corilagin[28], rosmarinic acid[29], and licorice polyphenols[30]) at concentrations of $10^{-6}-10^{-3}$M, while the formation of cyclooxygenase products (HHT, thromboxane B_2, and 6-keto-$PGF_{1\alpha}$) was generally not inhibited appreciably at these concentrations.

9. RADICAL-SCAVENGING ACTIVITY OF POLYPHENOLS ON DPPH

The radical-scavenging activities of polyphenols on DPPH, evaluated by colorimetry, are shown in Table 6. The activities of the polyphenols in the category of tannin showed potent inhibitory activities, while those of the polyphenols of small molecule, except EGCG, ECG and EGC, were generally low[31].

The generation of free radicals from these polyphenols in the experiment mentioned above were also proved by isolating the products of radical-coupling between two free radicals.

Table 8. Effects of polyphenols on superoxide radical

Polyphenols	Molecular weight	Number of phenolic OH	EC_{50} (μM)
(Hydrolyzable tannins)			
Galloylglucoses			
1,2,6-Tri-*O*-galloylglucose	636.5	9	4.1
1,2,3,6-Tetra-*O*-galloylglucose	788.6	12	2.8
1,2,3,4,6-Penta-*O*-galloylglucose	940.7	15	3.4
Ellagitannin monomers			
Pedunculagin	784.6	12	2.8
Casuarictin	936.7	15	2.8
Tellimagrandin II	938.7	15	3.1
Rugosin A	1106.8	17	2.7
Furosinin	816.6	4	7.5
Geraniin	952.7	11	3.0
Isoterchebin	954.7	11	3.1
Ellagitannins dimers			
Cornusiin A	1571.1	23	1.6
Rugosin D	1875.3	29	1.7
Coriariin A	1875.3	29	1.3
(Condensed tannins and their units)			
EC	290.3	4	23
EC dimer	578.5	8	14
EC trimer	866.8	12	8.8
EGC	306.3	5	1.6
EGCG	458.4	8	1.8
ECG	442.4	5	4.8
ECG dimer	882.7	10	5.0
ECG trimer	1323.1	15	2.6
ECG tetramer	1763.5	20	2.1
(Small polyphenols and others)			
Gallic acid	170.1	3	6.8
Methyl gallate	184.1	3	6.1
Ellagic acid	302.2	4	19
Ascorbic acid	176.1	–	32

Table 9. Effect of licorice phenols and quercetin on superoxide generated and detected in three systems

	Xanthine-XOD-NBT	PMS-NADH-NBT	Xanthin-XOD-cytochrome C (μM,IC_{50})
Licochalcone B	1.6	3.6	1.9
Glycyrrhisoflavone	4.2	10.5	8.0
Glisoflavone	14	20	15
Isoangustone A	20	61	11
Quercetin	5.0	4.0	3.0
Allopurinol	2.9	inactive	10.5

In this experiment, alkyl gallates having various length of alkyl group were mixed with DPPH radical in the solution, and the dimers produced by the coupling shown in Chart 6 were isolated by column chromatography. Among the alkyl gallates, *n*-propyl gallate yielded the dimer in the highest yield (41%), and the other alkyl gallates also gave the radical-coupling products in appreciable yields, showing that the free radicals were generated from alkyl gallates in the presence of DPPH, and these free radicals of rather small size easily coupled with each other[31]. The lability of the free radicals generated from the polyphenols of small molecule, exemplified by gallic acid and its derivatives, was shown by the ESR measurement of gallic acid, which exhibited the signals of HHDP radical producible by the mutual coupling of gallic acid radicals, without showing the free radical signal of gallic acid monomer. The isolation of the products of radical coupling thus showed chemically the lability of small polyphenol radicals[31] compared with the stabler free radicals from the polyphenols of larger molecule, which exhibited the ESR signals of monomer only, and did not give any product from the radical coupling[25].

10. INHIBITION OF XOD AND MAO

Polyphenols isolated from various plants (in Charts 1,2,4,5) also exhibited inhibitory effects on XOD, as shown by the IC_{50} values in Table 7[22]. Marked differences in the inhibition potency are observed in this table, and comparison of the relative inhibitory effects on XOD with the relative binding activities to protein in this table suggests that the inhibition of XOD by these polyphenols is not based on the "nonspecific binding to protein". The binding of polyphenols to xanthin oxidase therefore is not regarded as the main factor of the inhibitory activity on XOD[22]. Potent inhibitory effects of these polyophenols were observed on MAO, too[23].

11. INHIBITION OF SUPEROXIDE

The radical-scavenging mechanism of the inhibitory activity of polyphenols on XOD has been proved based on the data in Table 8 which were obtained by the following experiment. In this experiment, the 5,5-dimethyl-1-pyrroline-1-oxide (DMPO) adduct of the superoxide anion radical generated from hypoxanthine-xanthine oxidase system was measured by the ESR spectrometry. The scavenging effect of the polyphenols was ex-

Table 10. Effects of oral administration of geraniin on rat serum lipids

	Control	Treated with peroxidized oil	Geraniin (50mg/kg)	Geraniin (100mg/kg)
TG(mg/dl)	89.4	111.5	77.7	60.3
HDL-ch(mg/dl)	47.4	41.6	44.6	42.2
Atherogenic Index	0.90	1.63	0.80	0.44
LPO(MDA, nmol/ml)	3.40	6.35	4.58	4.26
FFA(meq/l)	0.182	1.35	1.02	0.81
TG(mg/dl)	112.5	351.7	166.1	130.5
GOT(Karmen Unit)	83.4	590.4	316.0	314.0
GPT(Karmen Unit)	32.2	436.8	259.0	225.0

pressed in terms of EC_{50} which is the concentration of polyphenols required to give a 50% decrease in the intensity of the signal of the DMPO adduct[32].

As the result, all the polyphenols tested in this experiment inhibited the appearance of the signal of the DMPO adduct in a dose-dependent manner. These ESR spectra showed the signal of DMPO adduct of *C*-centered radical, together with the signal of a hydrogen adduct of DMPO (DMPO-H), and a decrease in the intensity of the signal of the DMPO adduct of the superoxide anion radical. The appearance of the signal of the adduct of the *C*-centered radical suggested occurrence of the scavenging of superoxide anion radical by polyphenols through the formation of phenoxy radicals from the polyphenols[32].

The mechanism of the inhibition of XOD activity by these polyphenols, due to their radical-scavenging activity on superoxide, was further supported by the following experiment. In the experiment summarized by the data in Table 9, generation and detection of superoxide combined in three ways, and its suppression by polyphenols, were carried out to inspect whether the suppressive effect of each compound is reproducible regardless of the methods generating and detecting the superoxide radical. Four licorice phenolics, licochalcone B, glycyrrhisoflavone, glisoflavone and isoangustone A (structures in Chart 4), which showed appreciable inhibitory effect on the superoxide generated by the hypoxanthin-xanthinoxidase system, and quercetin as an active example of flavonol, were used as the inhibitors and their effects were compared with that of allopurinol, a clinically used medicine.

As shown in Table 9, all of these licorice phenolics appreciably suppressed the superoxide generated by three different methods. The differences in the methods of generation and detection of superoxide, either enzymatic or non-enzymatic, gave no marked difference of the result. This result therefore shows that the inhibition of XOD by the licorice polyphenols is attributable to direct suppressive activity on the superoxide radical generated in these systems.

The suppressive activities of these four polyphenols were higher than those of many other licorice phenolics hitherto isolated, and licochalcone B showed the most potent inhibition among them in the three experimental combinations. This is attributable to the presence of *ortho*-dihydroxy structures in their molecules, which allows generation of stabler free radicals than those of the phenolics having isolated hydroxyl group.

12. EFFECTS OF POLYPHENOLS BY ORAL ADMINISTRATION

12.1. Inhibition of Lipid Peroxidation in Serum, by Oral Administration of Geraniin

The antioxidant effects of some of these polyphenols were also exhibited by oral administration. The lipid peroxide level in the rat serum, which was raised by feeding peroxidized corn oil, was significantly lowered by the oral administration of geraniin as shown in Table 10[33]. Further experimental work will have to be carried out to see the correlations among this *in vivo* test and the *in vitro* experiments referred above.

12.2. Oral Administration of EGCG

EGCG was found to inhibit tumor promotion in the gastrointestinal tract in a model system of mouse duodenal carcinogenesis with ENNG. In this experiment ENNG was given to mice for four weeks, and then a half of treated mice were given 0.005% aqueous solution of EGCG, and the other half of the mice were given tap water. The percentage of tumor-bearing mice of the group treated with ENNG plus EGCG was 20% at week 16 after ENNG treatment, while that of the control group was 63%. Average number of tumors per mouse in the EGCG-treated group was only a quarter of the control group. Parallel results were obtained upon repetition of this experiment[34].

The participation of the activity of EGCG suppressing active-oxygen to this *in vivo* inhibitory effect of tumor promotion will be also one of the problems to be studied further concerning the activities of polyphenols.

13. CONCLUSION

It must be emphasized that there are a variety of chemical structures and difference of activities found for the plant polyphenols taken in by human beings. A large number of these compounds distribute widely, and accumulate in large amount in various plants. They could be potent antioxidants although we may often be taking them in without recognizing their presence. The significance of these plant polyphenols as the antioxidants in our lives can be summerized as follows:

1. Wide distribution in plants used as food and medicine.
2. Large variety in their chemical structures
3. Accumulation in large amount in plants
4. Lower toxicity (particularly upon oral administration) than that of many other types of natural organic compounds, and also than that imagined for plant polyphenols in the old time before the isolation and characetrization of each compound were successfully carried out.

ACKNOWLEDGMENTS

The author is deeply indebted to the help and co-operation given by eminent scientists whose names are in the references.

REFERENCES

1. Okuda, T., Yoshida, T. and Hatano, T. (1995) Hydrolyzable tannins and related polyphenols. In : W. Herz (ed), Progress in the chemistry of natural organic products, Vol. 66, pp.1- 117, Springer-Verlag, Wien New York.
2. Okuda, T., Yoshida, T and Hatano, T. (1991) Chemistry and biological activity of tannins in medicinal plants. In : H. Wagner and N. R. Farnsworth (eds), Economic and medicinal plant research, pp.129–165, Academic Press, London.
3. Okuda, T. (1995) Tannins, a new family of bio-active natural organic compounds. *Yakugaku Zasshi* 115:81–100.
4. Okuda, T. (1993) Natural polyphenols as antioxidants and their potential use in cancer. In : A. Scalbert (ed), Polyphenolic phenona, pp.221–235, INRA, Paris.
5. Okuda, T. (1994) Oriental medicine. In : E. Niki, H. Shimazaki and M. Mino (eds), Antioxidants, free radicals and biological defense, pp.263–275, Japan Scientific Societies Press, Tokyo.
6. Okuda, T. Mori, K. and Hatano, T. (1985) Relatioship of the structures of tannins to the binding activities with hemoglobin and methylene blue. *Chem. Pharm. Bull.* 33:1424- 1433.
7. Hatano, T., Kagawa, H., Yasuhara, T. and Okuda, T. (1988) *Chem. Pharm. Bull.* 36:2090–2097.
8. Viriot, C., Scalbert, A., Lapierre, C. and Moutounet M. (1992) European oak wood polyphenols and cognac ageing in oak barrels. Proceedings of Groupe Polyphenols, 16:242–245.

9. Okuda, T., Yoshida, T. and Hatano, T. (1989) Ellagitannins as active constituents of medicinal plants. *Planta Medica* 55: 117–234.

10. Okuda, T., Yoshida, T. and Hatano, T. (1980) Equilibrated stereostructures of hydrated geraniin and mallotusinic acid. *Tetrahedron Letters*, 21:2561–2564.

11. Okuda, T., Yoshida, T. and Hatano, T. (1982) Constituents of *Geranium thunbergii* Sieb. et Zucc. Part 12. *J. Chem. Soc. Perkin Trans I*:9–14.

12. Okuda, T., Yoshida, T. and Hatano, T. (1993) Classification of oligomeric hydrolyzable tannins and specificity of their occurrence in plants. *Phytochemistry* 32:507–521.

13. Okuda, T., Yoshida, T., Kuwahara, M., Memon, M. U. and Shingu, T. (1982) Agrimoniin and potentillin, an ellagitannin dimer and monomer having an a-glucose core. *J. Chem. Soc. Chem. Commun.*:163–164.

14. Hatano, T., Yasuhara, T., Matsuda, M., Yazaki, K. Yoshida, T. and Okuda, T. (1990) Oenothein B, a dimeric hydrolysable tannin with macrocyclic structure, and accompanying tannins from *Oenothera erythrosepala*. *J. Chem. Soc. Perkin Trans.*:2735–2743.

15. Okuda, T., Hatano T., Agata, I., Nishibe, S. and Kimura, K. (1986) Tannins in *Artemisia montana*, *A. princeps* and related species of plant. *Yakugaku Zasshi* 106:894–899.

16. Cavallini, L., Bindoli, A. and Siliprandi, N. (1978) Comparative evaluation of antiperoxidative action of silymarin and other flavonoids. *Pharmacol. Res. Commun.* 10:133–136.

17. Okuda, T., Hatano, T., Agata, I. and Nishibe, S. (1986) The components of tannic activities in Labiatae plants, I. *Yakugaku Zasshi* 106:1108–1111.

18. R. Hegnauer (1966) Chemotaxonomie der Pflanzen, Bd. 4, pp. 327–328, Birkhauser Verlag, Basel.

19. Agata, I., Kusakabe, H., Hatano, T., Nishibe, S. and Okuda, T. (1993) Melitric acids A and B, new trimeric caffeic acid derivatives from *Melissa officinalis*. *Chem. Pharm. Bull.* 41:1608–1611.

20. Agata, I., Hatano, T., Nishibe, S. and Okuda, T. (1989) A tetrameric derivative of caffeic acid from *Rabdosia japonica*. *Phytochemistry* 28:2447–2450.

21. Hatano, T., Fukuda, T., Liu, Y.-Z., Noro, T. and Okuda, T. (1991) Phenolic constituents of licorice, IV. *Yakugaku Zasshi* 111:311–321.

22. Hatano, T., Yasuhara, T., Fukuda, T., Noro, T. and Okuda, T. (1989) Phenolic Constituents of licorice II, Structures of licopyranocoumarin, licoarylcoumarin and glisoflavone, and inhibitory effects of licorice phenolics on xanthine oxidase. *Chem. Pharm. Bull.* 37:3005–3009.

23. Hatano, T., Fukuda, T., Miyase, T., Noro, T. and Okuda, T. (1991) Phenolic constituents of licorice, III. *Chem. Pharm. Bull.* 39:1238–1243.

24. Fujita, Y., Uehara, I., Morimoto, Y., Nakashima, M., Hatano, T. and Okuda, T. (1988) Studies on inhibition mechanism of autoxidation by tannins and flavonoids, II. *Yakugaku Zasshi* 108:129–135.

25. Fujita, Y., Komagoe, K., Niwa, Y., Uehara, I., Hara, R. , Mori H. , Okuda, T. and Yoshida, T. (1988) Studies on inhibition mechanism of autoxidation by tannins and flavonoids, III. *Yakugaku Zasshi* 108:528–537.

26. Okuda, T., Kimura, Y., Yoshida, T., Hatano, T., Okuda, H. and Arichi, S. (1983) Studies on the activities of tannins and related compounds from medicinal plants and drugs, I. *Chem. Pharm. Bull.* 31:1625–1631.

27. Hikino, H., Kiso, Y., Hatano, T., Yoshida, T. and Okuda, T. (1985) Antihepatotoxic actions of tannins. *J. Ethnopharma- cology*, 14:19–29.

28. Kimura, Y., Okuda, H., Okuda, T., Hatano, T., Agata, I. and Arichi, S. (1984) Studies on the activities of tannins and related compounds, V. *Planta Medica*:473- 477.

29. Kimura, Y., Okuda, H., Okuda, T., Hatano, T. and Arichi, S. (1987) Studies on the activities of tannins and related compounds, X. *J. Nat. Prod.* 50:392–399.

30. Kimura, Y. Okuda, H., Okuda, T. and Arichi, S. (1988) Effects of chalcones isolated from licorice root on leukotriene biosynthesis in human polymorphonuclear neutrophils. *Phytotherapy Res.*, 2, 140–145.

31. Yoshida, T., Mori, K., Hatano, T., Okumura, T., Uehara, I., Komagoe, K., Fujita, Y. and Okuda, T. (1989) Studies on inhibition mechanism of autoxidation by tannins and flavonoids, V. *Chem. Pharm. Bull.* 37:1919–1921.

32. Hatano, T., Edamatsu, R., Hiramatsu, M., Mori, A., Fujita, Y., Yasuhara, T., Yoshida, T. and Okuda, T. (1989) Effects of the interaction of tannins with co-existing substances. *Chem. Pharm. Bull.* 37:2016–2021.

33. Kimura, Y., Okuda, H., Mori, K., Okuda, T. and Arichi, S. (1984) Studies on the activities of tannins and related compounds from medicinal plants and drugs, IV. *Chem. Pharm. Bull.* 32:1866–1871.

34. Fujita, Y., Yamane, T., Tanaka, M., Kuwata, K., Okuzumi, J., Takahashi, T., Fujiki, H. and Okuda, T. (1989) Inhibitory effect of (-)-epigallocatechin gallate on carcinogenesis with *N*-ethyl-*N'*-nitro-*N*-nitrosoguanidine in mouse duodenum. *Jpn. J. Cancer Res.* 80:503–505.

5

ANTIOXIDANTS IN TEA AND THEIR PHYSIOLOGICAL FUNCTIONS

Yukihiko Hara

Food Research Labs
Mitsui Norin Co., Ltd.
Fujieda, Fujieda City 426, Japan

The dried young leaves of the plant Camellia sinesis and the infusion of these leaves is called tea. Tea is characterized by its pungency as well as its aroma. The compounds responsible for the pungency of tea are called tea polyphenols. In this article, the extraction, characteristics, and antioxidative properties of tea polyphenols in various systems are described.

1. METHOD OF EXTRACTING TEA POLYPHENOLS

An analysis of the constituents of young tea leaves (Assam variety) is shown in Table 1. As seen in the table, tea polyphenols constitute the major part of the water soluble fraction of tea leaves. In their natural state, tea polyphenols are in the form of catechins. Processes of green tea manufacture do not effect these catechins which remain unchanged.

Catechin fractions (Green Tea Catechin: GTC) and individual catechins are extracted by hot water from green tea. The method of extracting catechin fractions from green tea is shown in Fig.1. GTC thus extracted is made up principally of five individual catechins. The method of separation of individual catechins by HPLC is shown in Fig. 2. Purified catechins are needle-like crystals which are either transparent or whitish in color. The structural formulae of catechins are shown in Fig. 3. An example of the composition of catechins obtained from Japanese green tea is shown in Table 2. In black tea manufacture, tea leaves are rolled and crushed, followed by a process called fermentation in which rolled leaves are spread out and stacked in order to facilitate enzymatic oxidation. During these rolling and fermentation processes, catechins are oxidized to form theaflavins or thearubigins which are mostly dimeric catechins and which are responsible for the reddish brown color of black tea infusion.

The separation process of theaflavins and thearubigin fraction is shown in Fig. 4. Although theaflavin fraction constitutes only about 1% of black tea (dry weight), it represents the bright, red color of black tea. Thearubigin fraction is assumed to be made up of

Table 1. Composition of fresh tea flush—Assam variety*

Components	Amount found in flush (% dry wt.)
Substances soluble in hot water	
Flavanols	17 - 30
Flavonols and flavonol glycisides	3 - 4
Leucoanthocyanins	2 - 3
Polyphenolic acids and depsides	~ 5
Total polyphenols	~ 30
Caffeine	3 - 4
Amino acids	~ 4
Simple carbohydrates	~ 4
Organic acids	~ 0.5
Substances partially soluble in hot water	
Polysaccharides	~ 13
Proteins	~ 15
Ash	~ 5
Substances insoluble in water	
Cellulose	~ 7
Lignin	~ 6
Lipids	~ 3
Pigments	~ 0.5
Volatile substances	0.01 - 0.02

*Adapted from Millin and Rustidge (1967)

G.W.Sanderson: Recent Advances in Phytochemistry
(ed. by V.C.Runeckles and T.C.Tso), No.5, p.265 (1972),
Phytochemical Society of North America

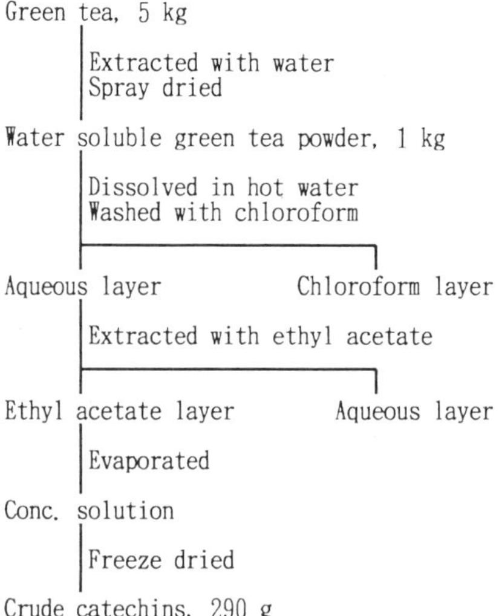

Figure 1. Preparation of "Green Tea Catechin (GTC)"

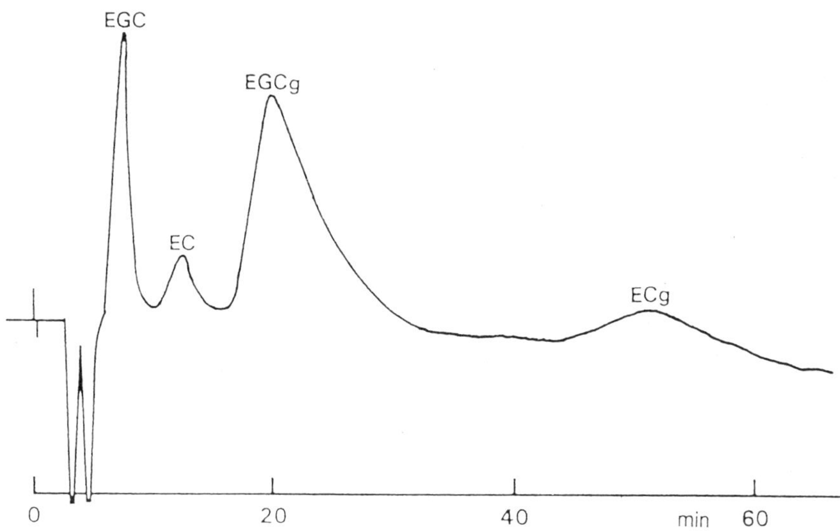

Figure 2. High Performance Liquid Chromatogram (HPLC) of Green Tea Catechin. Instrument, Waters Prep LC/System 500A; detector, RID; column, Waters Prep PAC-500/C18 (5cmϕ×30cm); mobile phase, tetrahydrofuran/aceton/H$_2$0 (10/12/78); flow rate, 150ml/min.

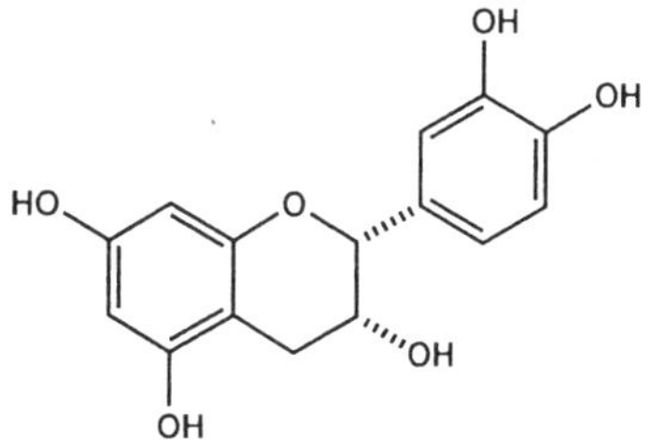

(-)-Epicatechin   (EC)

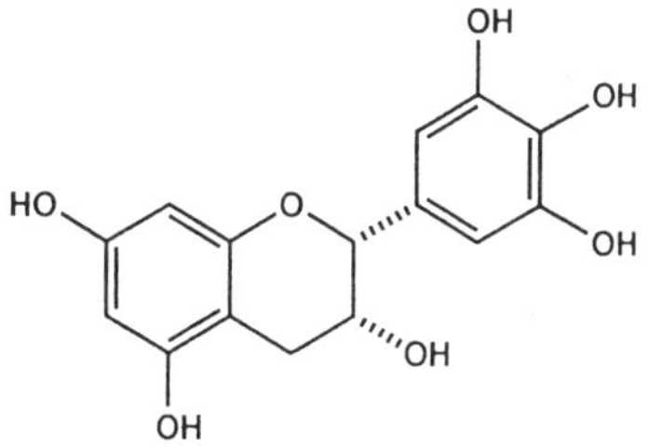

(-)-Epigallocatechin   (EGC)

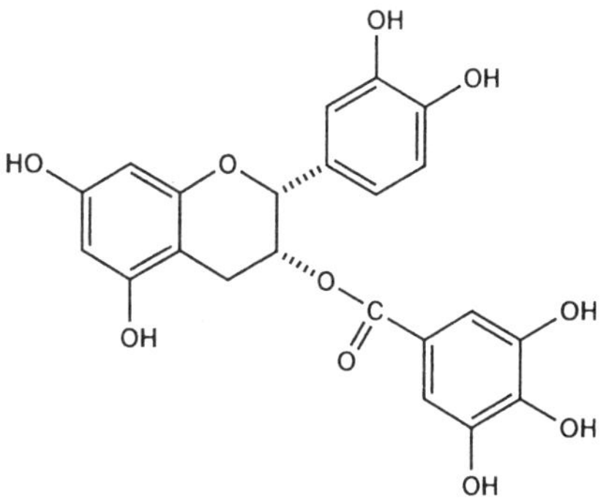

(-)-Epicatechin gallate (ECg)

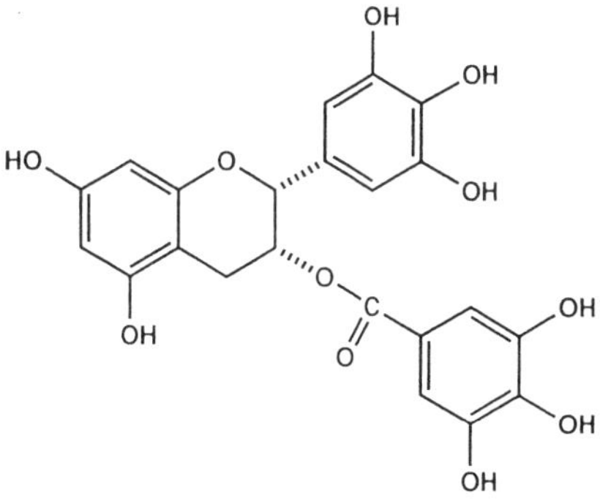

(-)-Epigallocatechin gallate (EGCg)

Figure 3. Structural formulae of catechins.

Table 2. Composition of "Green Tea Catechin" in green tea

Catechins	Absolute %	Relative %
(+)-Gallocatechin (GC)	1.44	1.6
(-)-Epigallocatechin (EGC)	17.57	19.3
(-)-Epicatechin (EC)	5.81	6.4
(-)-Epigallocatechin gallate (EGCg)	53.90	59.1
(-)-Epicatechin gallate (ECg)	12.51 / 91.23	13.7 / 100

various unidentified catechin dimers and oligomers, and they make the brown, red color of black tea infusion.

The separation process of individual theaflavins from crude theavins is shown in Fig. 5. The composition of crude theaflavins is shown in Table 3. Purified theaflavins form reddish-orange column-shaped crystals. The structural formulae of theaflavins are shown in Fig. 6. Catechins and theaflavins were proven to have very potent antioxidative and radical scavenging action in various systems. The experimental results are described in the following chapters.

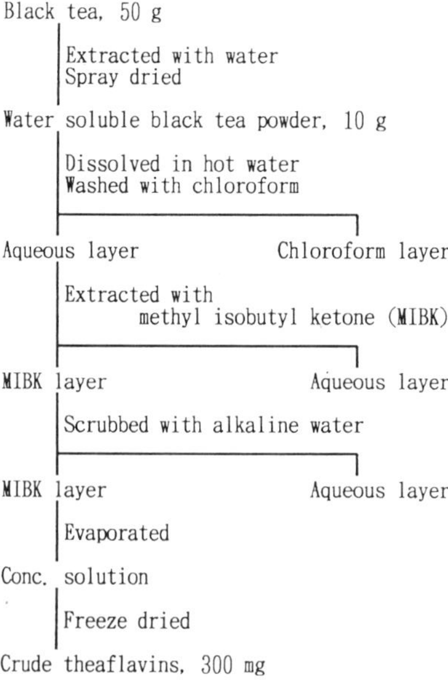

Figure 4. Preparation of "crude theaflavins."

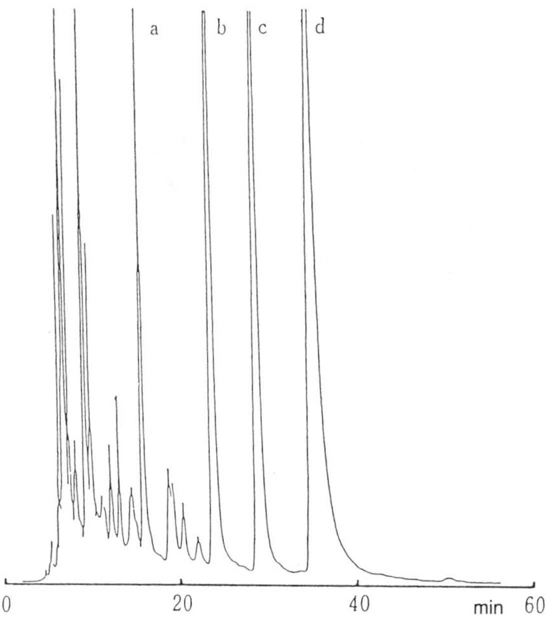

Figure 5. High performance liquid chromatogram of crude theaflavins. Instruments: Shimadzu LC4A, column: YMC A-321ODS (6cm×15cm), mobile phase: acetonitril/etylacetate/0.05% phosphoric acid (21/3/76), flow rate: 1 ml/min.

2. RADICAL SCAVENGING ACTION OF TEA CATECHINS

There have been various reports on the radical scavenging action of tea polyphenols(1)(2)(3). We examined the reaction of catechins with DPPH radical(4). Catechins were dissolved in ethanol at pH 4 with DPPH radical and the intensity of pulses were recorded by ESR. As shown in Fig. 7, DPPH radical scavenging activity of catechins was in the decreasing order of EGCg>EGC≥ECg>EC≥C. Since catechins are known to form quiones more easily in an alkali environment, the radical scavenging ability of (+)-cate-

Table 3. Composition of "crude theaflavins" in black tea

Theaflavins	Absolute %	Relative %
Theaflavin (TF1)	13.35	15.8
Theaflavin monogallate A (TF2A)	18.92	22.4
Theaflavin monogallate B (TF2B)	18.64	22.0
Theaflavin digallate (TF3)	33.49 ⁄ 84.40	39.7 ⁄ 100

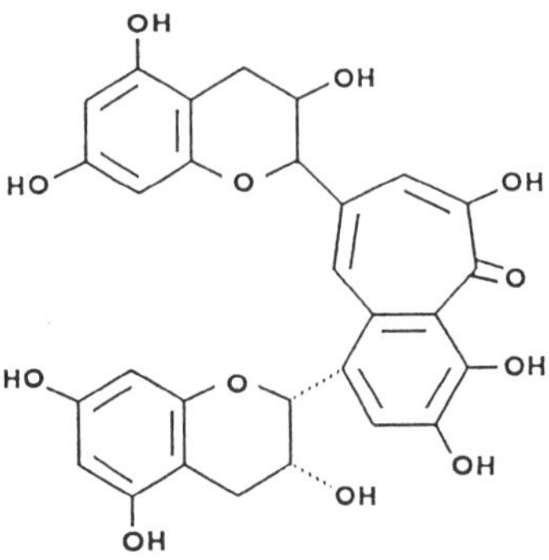

Theaflavin (T F 1)

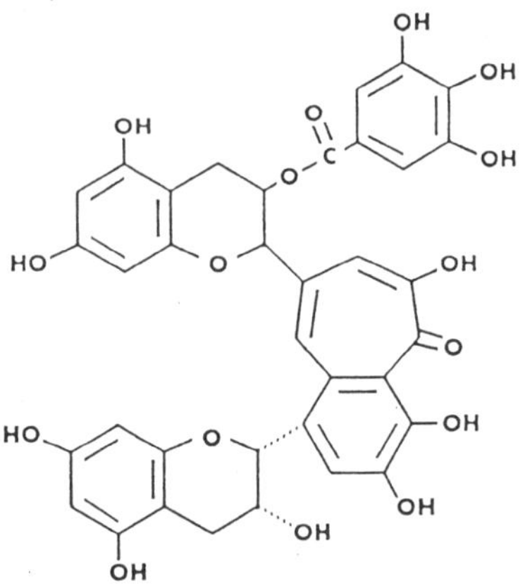

Theaflavin monogallate A
(T F 2 A)

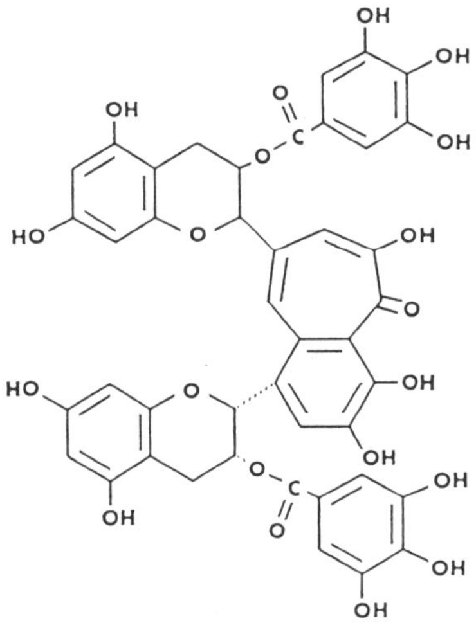

Theaflavin digallate (T F 3)

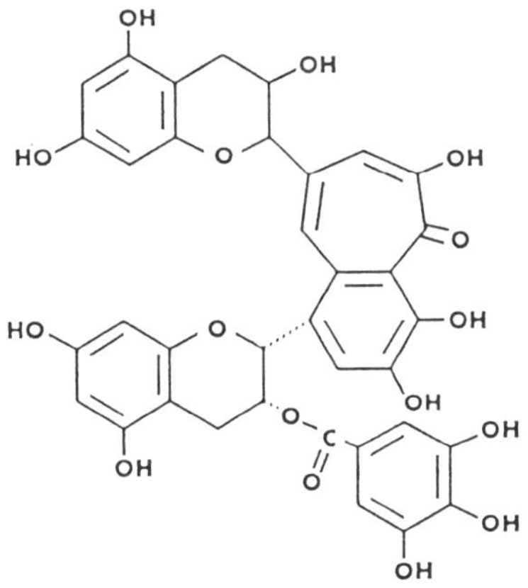

Theaflavin monogallate B
(T F 2 B)

Figure 6. Structural formulae of theaflavins.

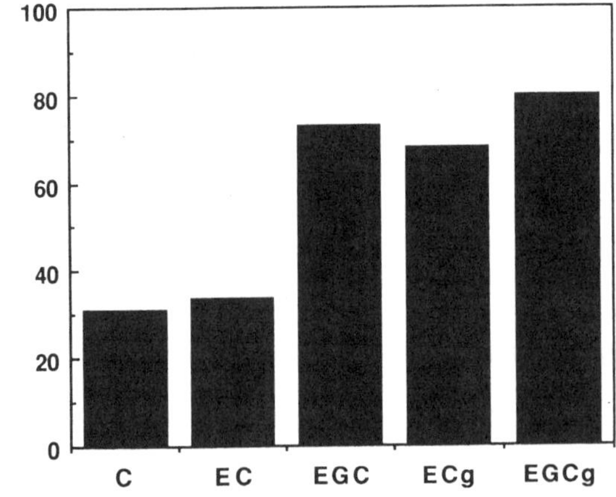

Figure 7. DPPH radical scavenging ability of green tea catechin.

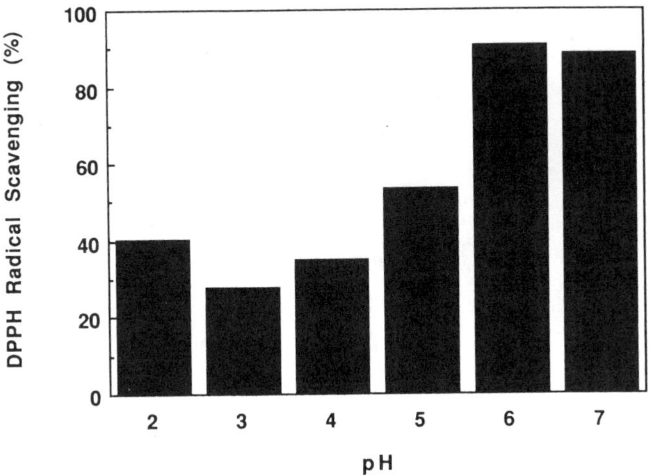

Figure 8. Effects of pH on radical scavenging ability of (+)-catechin.

chin was measured in the same system as above but at different pH. The results show that as pH increases from 3 to 7, the radical scavenging action also increases (Fig. 8). At this point in our research, we assume the molecular mechanism of radical scavenging and antioxidative actions of tea catechins to be as follows(Fig. 9): Catechins donate hydrogen from a hydroxyl group to radicals and so terminate the radical chain reaction. A phenoxyradical thus made will scavenge another radical. Finally, they make quione structures and become stable.

In this assumption, the presence of orthohydroxy groups in the B ring is essential for potent radical scavenging action. We are confirming this fact by glycosylating OH groups at various positions(5). At the same time, it is assumed from the above results that the tendency of the OH group on the B ring to ionize will influence the potency of catechin radical scavenging ability. These details are currently under study.

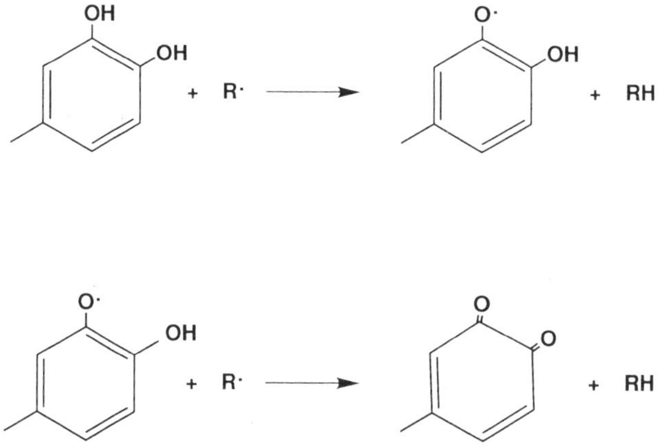

Figure 9.

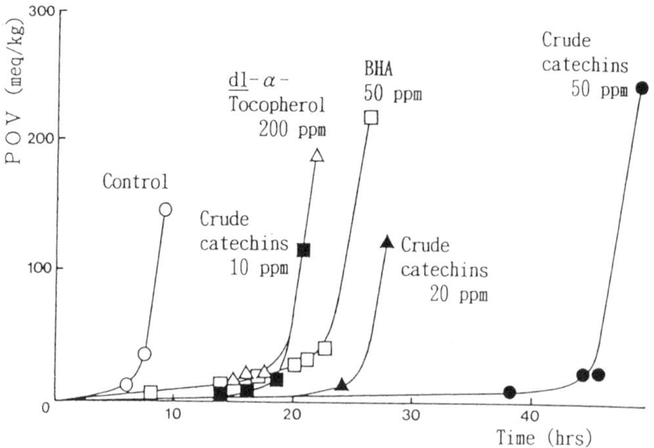

Figure 10. Antioxidative activities of GTC on lard (AOM at 97.8°C).

3. ANTIOXIDATIVE ACTION OF TEA CATECHINS IN EDIBLE FATS AND OILS

Oxidation of lard was measured by Active Oxygen Method(AOM)(6). In this method, lard was put in a glass cylinder and heated in a silicon oil bath at the consistent temperature 97.8° with a continuous bubbling of air inside the cylinder to facilitate peroxidation. After a few hours of this induction period, rancidity of the lard began and as oxidation progressed peroxide value (POV) increased. After another few hours POV increased exponentially as shown in Fig.10. The addition of GTC to the lard prolonged the induction period markedly by suppressing the peroxidation of the lard. In this system, the antioxida-

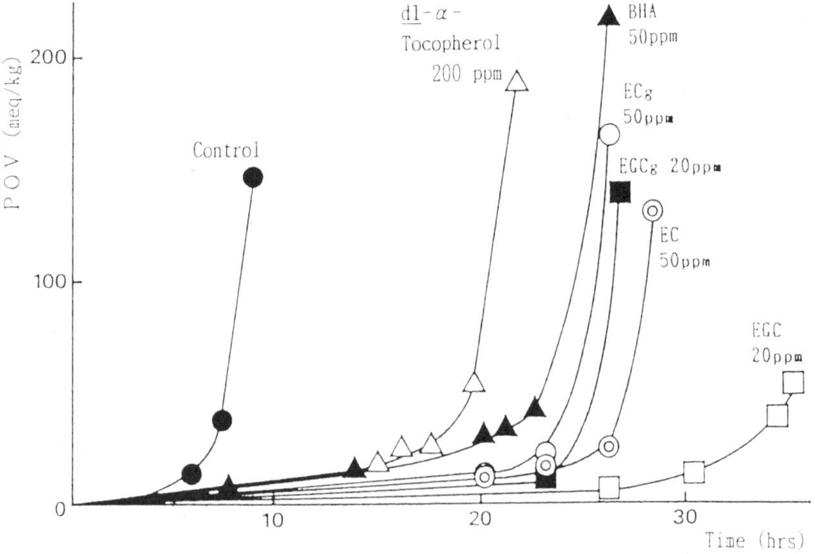

Figure 11. Antioxidative activities of green tea catechin on lard (AOM at 97.8 °C)

tivity of GTC increased dose-dependently at 10, 20 and 50ppm. In suppressing peroxidation, the addition of just 10ppm of GTC equaled or surpassed 200ppm of *dl*—tocopherol or 50ppm of BHA as shown in the figure. With regard to the antioxidativity of individual catechins, galloyl catechins(EGCg and EGC) were more potent than catechol catechins(ECg and EC) as calculated by the results in Fig.11. In the same AOM system, docosahexanoic acid(DHA), a functional polyunsaturated fatty acid in fish oil, was tested with catechins. As shown in Fig.12, the addition of tea catechins suppressed the oxidation of DHA. In the same manner, antioxidativity was proved in various other edible oils.

4. ANTIOXIDATIVE ACTION OF THEAFLAVINS ON ERYTHROCYTE MEMBRANE LIPIDS(7)

As a functional lipid in the body, erythrocyte membrane ghost was separated for the *in vitro* oxidation test. Rabbit's erythrocyte ghost was subjected to oxidation by mixing it with *tertial* butyl hydroperoxide(BHP) and incubating it at 37° for 30 min. The degree of oxidation was determined by the coloring of the above solution with the addition of thiobarbituric acid(TBA). Antioxidants were mixed with the lipid before the addition of BHP. As shown in Fig. 13, theaflavins were more potent in suppressing the oxidation of erythrocyte ghost than α-tocopherol or propyl gallate in the same concentrations (final concentration: 25μl). Also in the same system, catechins were shown to exert similar antioxidativity as theaflavins.

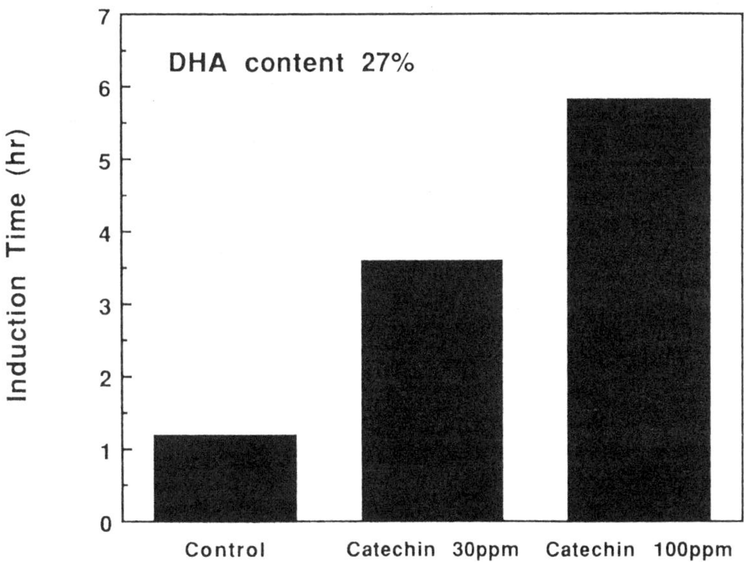

Figure 12. Antioxidative effects of green tea catechin on fish oil with high DHA content.

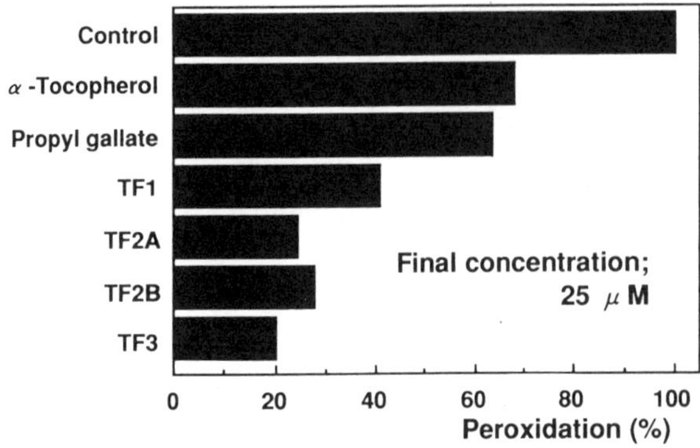

Figure 13. Antioxidative activity of theaflavins determined by rabbit erythrocyte ghost system.

5. INFLUENCE OF CATECHIN FEEDING ON THE LEVELS OF α-TOCOPHEROL, TBARS IN PLASMA AND ERYTHROCYTES(8)

In order to investigate the influence of catechin feeding in a diet rich with saturated or polyunsaturated fatty acid, male Wistar rats 5 weeks of age were fed on the following four diets for 31 days:(1)30% palm oil diet; (2) 30% palm oil diet containing 1% tea catechin; (3) 30% perilla oil diet; (4) 30% perilla oil diet containing 1% tea catechin. Each group consisted of six rats and all rats were caged individually. The composition of the diets in each group is shown in Table 4, and the composition of fatty acids in Palm and Per-

Table 4. Composition of high fat diets

Ingredient	Palm oil group		Perilla oil group	
	Control	Catechin	Control	Catechin
Corn starch	28.9 (%)	28.9 (%)	28.9 (%)	28.9 (%)
Sucrose	10.0	10.0	10.0	10.0
Casein	20.0	20.0	20.0	20.0
Palm oil	30.0	30.0	-	-
Perilla oil	-	-	30.0	30.0
Cellulose	5.0	5.0	5.0	5.0
Salt mixture [a]	4.0	4.0	4.0	4.0
Choline chloride	0.1	0.1	0.1	0.1
Vitamin mixture [a] (vitamin E free)	2.0	2.0	2.0	2.0
Tea catechin	-	1.0	-	1.0
α-Tocopherol [b]	3.7 mg	3.7 mg	4.3 mg	4.3 mg

[a] Salt mixture and vitamin mixture (vitamin E free) according to Harper were purchased from Oriental Kobo Kogyo Co..

[b] Taking into account the content of α-tocopherol in the palm and perilla oils, the final concentration of the tocopherol in the diets was adjusted to 6 mg/100 g.

Table 5. Fatty acid composition of palm and perilla oils

Fatty acid	Palm oil (%)	Perilla oil (%)
14 : 0	1.1	-
16 : 0	46.6	6.7
16 : 1	-	-
18 : 0	3.8	2.1
18 : 1	37.5	17.7
18 : 2	9.8	15.5
18 : 3	-	52.6
20 : 0	0.2	-

illa oils is shown in Table 5. To each diet, -tocopherol was added to give the same concentration of 6mg/100g, taking into account the intrinsic amount already contained in the oils. Food and water were fed *ad libitum*. At the end of the feeding period, rats were fasted overnight, anesthetized and blood was collected by heart puncture. The plasma and red blood cells were separated. During the feeding period, there were no significant differences among the groups tested in terms of either body weight gain or food intake. The effects of dietary tea catechins on the levels of both -tocopherol and TBA-reactive substances(TBARS) in the plasma and erythrocytes were examined as well as the plasma lipid levels.

The plasma lipid levels in the perilla oil fed rats were markedly lower than those in the palm oil rats, regardless of tea catechin supplementation (Table 6). As shown in Fig. 14, TBARS content in erythrocytes was not influenced very much by catechin supplementation as well as in the two oil groups. However, there was a big difference between TBARS in plasma between the two oil groups; perilla oil (polyunsaturated) groups showed much higher TBARS values than those of palm oil(saturated) groups. In perilla oil groups, tea catechin supplementation suppressed TBARS significantly.

Table 6. Concentration of cholesterol, phospholipids, and triglycerides in plasma

Lipids	Palm oil group		Perilla oil group	
	Control	Catechin	Control	Catechin
Cholesterol (mg/dl)	65.9 ± 2.7	66.7 ± 10.0	36.8 ± 2.9 [a]	24.7 ± 5.0 [a,b]
Phospholipids (mg/dl)	126.9 ± 11.4	121.1 ± 9.5	68.0 ± 14.9 [a]	61.7 ± 11.1 [a]
Triglycerides (mg/dl)	80.3 ± 10.7	67.3 ± 16.3	32.2 ± 8.0 [a]	26.3 ± 3.7 [a]

Values were expressed as mean ± SD.

[a] Statistically significant differences compared with palm oil group, p<0.01

[b] Statistically significant differences compared with perilla oil group, p<0.001

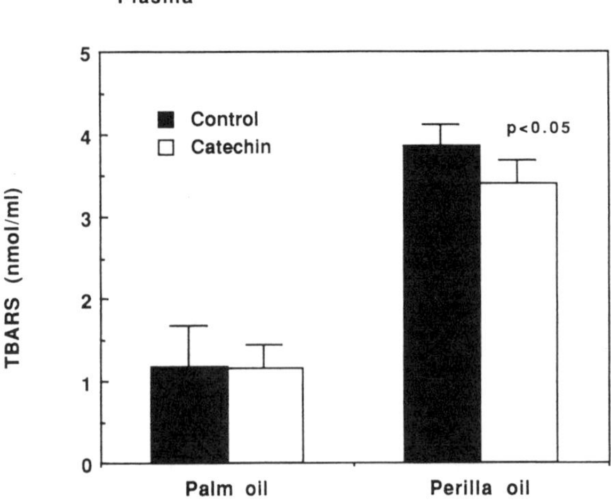

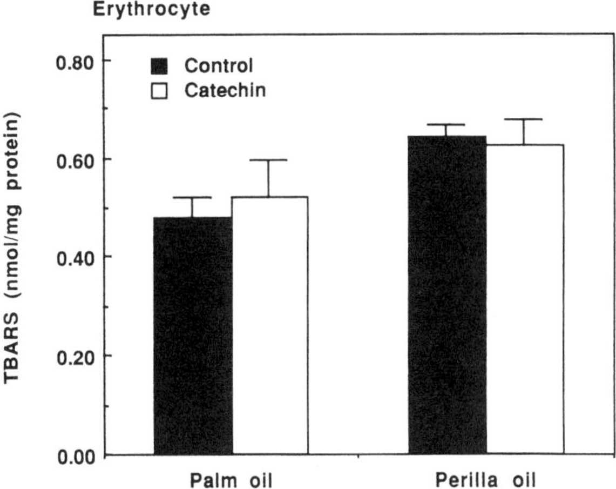

Figure 14. Effects of green tea catechin on lipid peroxide levels in plasma and erythrocyte.

The concentration of -tocopherol was examined in four groups and the results are shown in Fig. 15. It is apparent that for both the plasma and erythrocyte, the content of -tocopherol is much lower in the perilla oil fed group. Polyunsaturated oil(perilla oil) seems to consume more -tocopherol than saturated oil does because of its susceptibility to oxidation. Yet the supplementation of catechins inhibits the consumption of -tocopherol significantly in every case. These phenomena suggest that oxidative damage in the body can be ameliorated by the supplementation of tea catechins in the diet.

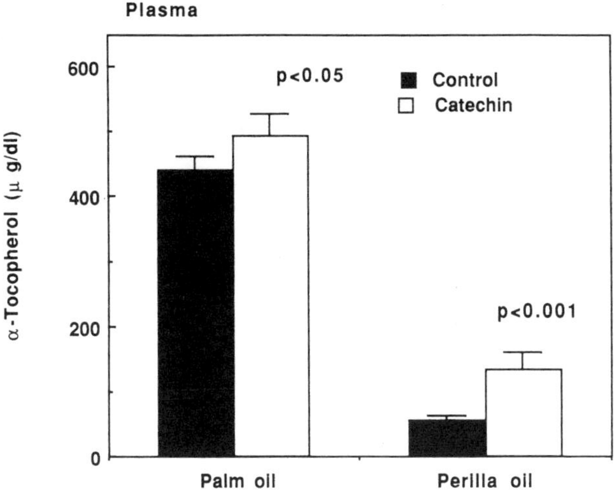

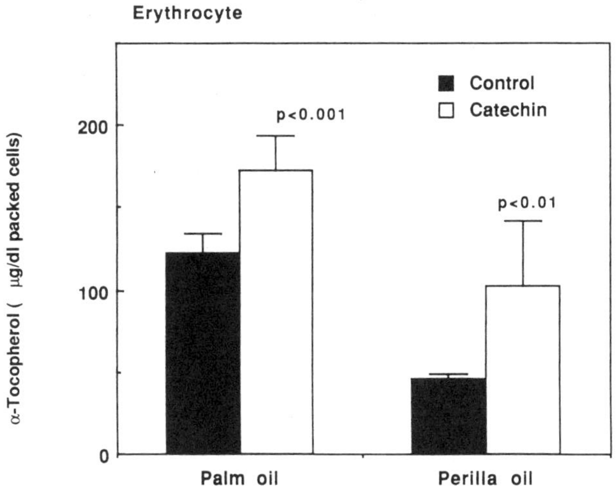

Figure 15. Effects of green tea catechin on α-tocopherol levels in plasma and erythrocyte.

6. EFFECTS OF LONG-TERM DIETARY SUPPLEMENTATION OF TEA POLYPHENOLS ON LIPID PEROXIDE LEVELS IN RATS(9)

An age-dependent increase of lipid peroxidation has been reported(10) to be closely associated with age-related diseases such as atherosclerosis, diabetes mellitus and the aging process. Sano *et al.* reported that liver and kidney slices from rats fed tea leaf powder(3% green tea or black tea added to the basal diet) for 50 days showed an increased resistance against lipid peroxidation after being treated by *t*-butyl hydroperoxide (BHP) or bromotrichloromethane. Similar research in the past has often been conducted under drastic conditions. For example, dosing with artificial chemicals over a short period of time or dosing with extreme amounts of certain nutrients as described in the previous chapter. In this experiment, rats were fed a moderate dosage of tea catechins(GTC at 0.5% and 1.0% in the normal diet) for a period of more than half the life span of rats i.e., from weaning at 3 weeks to full adulthood at 19 months, in order to clarify whether or not dietary antioxidants could ameliorate the condition of lipid peroxidation over a long period.

Firstly, the effects of dietary supplementation of GTC on growth, food intake and on the liver and kidney weights were examined. There were no statistically significant differences between body weights of a group of rats fed a normal diet(control group) and two groups of rats fed GTC-supplemented diets throughout the test period except at the 7–11 weeks stage. The body weight gain in GTC groups at 7–11 weeks was significantly lower than that of the control group. There were no significant differences in liver and kidney weights among the three groups. In Table 7, the weight of the liver and kidneys was shown at different GTC doses with time. The number in the parenthesis indicates the number of rats used.

Effects of dietary GTC supplementation on Thiobarbituric Acid Reactrive Substances(TBARS) in the plasma are shown in Fig. 16. The TBARS values in older rats(13 and 19 months old) were significantly higher than those of younger rats in every group regardless of diet. Similar trends showing this age dependent increase were observed with TBARS in the liver and kidneys. In Fig. 16, there were no statistically significant differences in TBARS values among the three groups although numerically GTC groups showed lower numerical values as compared to the control group. When the data of two age groups at 13 and 19 months were combined, these middle age groups showed significantly lower TBARS with GTC groups as compared with the control group. The combined TBARS data in the 1.0% GTC group was 11.8 ± 1.7 nmol/ml as against the combined control data of 15.3 ± 3.0.

Table 7. Effect of GTC feeding on liver and kidney weight of rats

Age	Dose of GTC (%)		
(months)	0	0.5	1.0
Liver weight (g)			
1-1.5	8.38±2.08(10)	7.73±1.76(10)	7.36±1.83(10)
2-3	14.04±2.30(7)	10.79±3.86(8)	11.30±2.78(6)
5-7	14.03±1.86(7)	13.48±0.58(7)	12.33±1.40(8)
13	12.07±1.66(3)	11.00±0.82(3)	11.33±1.00(3)
19	17.47±2.00(3)	16.56±0.12(3)	13.12±1.84(3)
Kidney weight (g)			
1-1.5	1.47±0.23(10)	1.42±0.26(10)	1.38±0.24(10)
2-3	2.13±0.33(7)	1.87±0.44(8)	2.11±0.55(7)
5-7	2.45±0.22(8)	2.40±0.34(7)	2.28±0.17(7)
13	2.49±0.13(3)	2.33±0.24(3)	2.33±0.15(3)
19	3.04±0.21(3)	2.77±0.13(3)	2.63±0.16(3)

Mean values±SD are given.The figure in each parenthesis indicates the number of rats used.

The contents of various lipids, such as tryglyceride(TG), total cholesterol(TC), phospholipid(PL) and non-esterified fatty acids(NEFA) in plasma of animals from weaning to 19 months of age are shown in Fig.17. The contents of TG, TC and PL in plasma of the control group at 19 months were significantly higher than all the other younger age groups(p<0.05). Dietary supplementation with GTC stimulated significant decreases in plasma lipids. Namely, the contents of TG and TC in the 1.0% GTC group were significantly lower than those of the control group at 19 months of age(p<0.05), and the contents of plasma TC and PL in the 0.5% and 1.0% groups were significantly lower than those in the control at 13 months of age(p<0.05). The contents of NEFA were not significantly different among the three dietary groups. We have separately confirmed in the 1.0% GTC group at 13–19 months of age a positive correlation between TBARS and TG, TC and PL levels of plasma.

It is noteworthy that long-term dietary supplementation of GTC could bring about a decrease in lipids and lipid peroxides in plasma. And it is suggested that the remarkable decreases of plasma lipid peroxides observed with dietary GTC supplementation is due, at least partly, to the antioxidant effects of GTC.

SUMMARY

Tea catechins were extracted from green tea and theaflavins, the dimeric compounds of catechins, were extracted from black tea. These polyphenolic compounds or a crude mixture of them were put in various oxidative systems in order to determine their antioxidative potency.

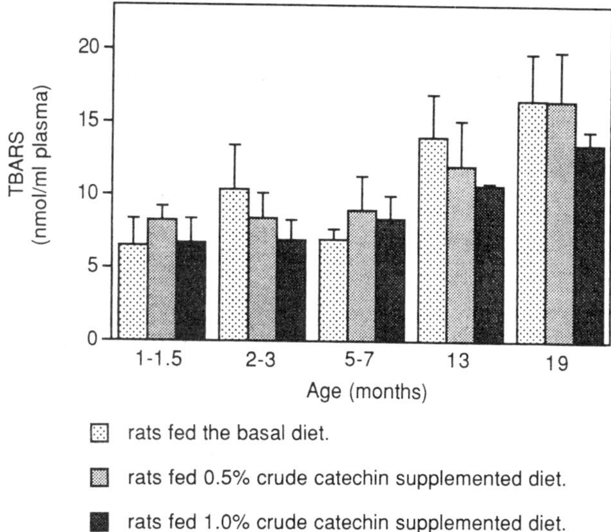

Figure 16. Effects of dietary supplementation of crude catechins on TBARS in the plsma. Mean values ±SD are given.

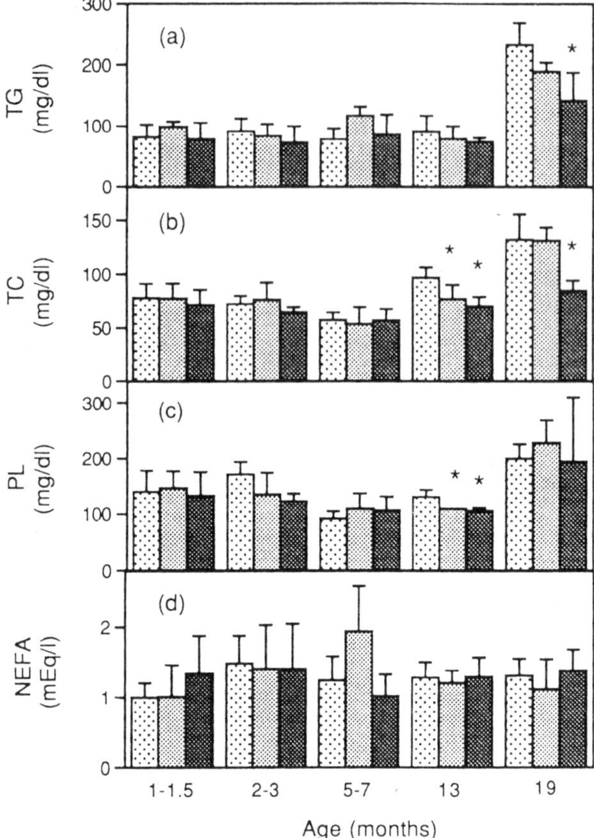

Figure 17. Effects of dietary supplementation of crude catechins on the content of plasma lipids of rats. (a), TG contents; (b), TC contents; (c), PL contents; (d), NEFA contents. Mean values ±SD are given. Significantly different from the values of rats fed the basal diet; *p<0.05.

Catechins were proved to have potent radical scavenging action against DPPH radical. A hypothesis on the mechanisms of radical scavenging of catechin was proposed. The suppression of peroxidation of edible oils e.g., lard, by tea catechins, was demonstrated. The effects were much more potent than α-tocopherol or BHA on a weight basis. In an *in vitro* erythrocyte membrane lipid system, theaflavins as well as catechins suppressed the peroxidation induced by *t*-BHP.

Very high doses of saturated or polyunsaturated fatty acids were included in the diet of rats and the influence of catechin feeding on lipids in plasma and erythrocyte was investigated. In the perilla oil(polyunsaturated oil) fed group, the accumulation of lipids in plasma was significantly suppressed by catechin feeding. Catechin feeding was proved to prevent the consumption of -tocopherol in plasma and erythrocyte.

Catechin feeding of 0.5% and 1.0% in a normal diet was carried out throughout the life-span of rats. At the 13–19 months old stage, the increase of lipids (TG, TC, PL) and TBARS in plasma was apparent. The addition of tea catechins significantly suppressed the increase of these lipids.

During this long term feeding, there were no significant differences in body weight, food intake, or liver and kidney weights between the control and test groups. From this data, it can be concluded that daily drinking of tea might be beneficial in reducing the accumulation of excessive lipids in the plasma and keep them from oxidative damage in humans. This data also confirms the safety of life long catechin consumption at a dose of 1% in the diet, which is equivalent to about 50 cups/day of ordinary strength green tea.

REFERENCES

1. Uchida, S., Edamatsu, R., Hiramatsu, M., Mori, A., Nonaka, G., Nishioka, I., Niwa, M. and Ozaki, M. (1987) Condensed tannins scavenge active oxygen free radicals. Med. Sci. Res. 15:831–832.
2. Zhao, B., Li, X., He, R., Cheng, S., and Wenjuan, X. (1989) Scavenging effect of extracts of green tea and natural antioxidants on active oxygen radicals. The Humana Press Inc.
3. Yoshida, T., Mori, K., Hatano, T., Okumura, T., Uehara, I., Komagoe, K., Fujita, Y. and Okuda, T. (1989) Studies on inhibition mechanism of autoxidation by tannins and flavanoids. Radical scavenging effects of tannins and related polyphenols on 1,1-Diphenyl-2-picrylhydrazyl radical. Chem. Pharm. Bull. 37(7) 1919–1921.
4. Nanjo, F., Hara, Y. Radical scavenging effects of tea catechins. (1995) Abstracts of International Symposium on Natural Antioxidants: Molecular Mechanisms and Health Effects.
5. Suzuki, M., Seto, R., Okushio, K., Sakai, M., Nanjo, F., Hara, Y., Bandai, T. and Sugimoto, T. (1994) Enzymatic synthesis of polyphenol glycoside. Nippon Nogeikagaku Kaishi, Vol. 68, No. 3, p.581.
6. Hara, Y. Prophylactic functions of tea polyphenols. (1994) In: C. Ho, (ed.), T, Osawa (ed.), M. Huang (ed.), R. Rosen (ed.), ACS Symposium Series 547, Food Phytochemicals for Cancer Prevention II, Teas, Spices, and Herbs.
7. Shiraki, M,. Hara, Y., Osawa, T., Kumon, H., Nakayama, T. and Kawakishi, S. (1994) Antioxidative and antimutagenic effects of theaflavins from black tea. Mutation Research, 323:29–34.
8. Nanjo, F., Honda, M., Okushio, K., Matsumoto, N., Ishigaki, F., Ishigami, T. and Hara, Y. (1993) Effects of dietary tea catechins on α-tocopherol levels, lipid peroxidation, and erythrocyte deformability in rats fed on high palm oil and perill oil diets. Biol. Pharm. Bull. 16(11):1156–1159.
9. Yoshino, K., Tomita, I., Sano, M., Oguni, I., Hara, Y. and Nakano, M. (1994) Effects of long-term dietary supplement of tea polyphenols on lipid peroxide levels in rats. Age, Vol. 17:79–85.
10. Sano, M., Takahashi, Y., Yoshino, K., Shimoi, K., Nakamura, Y., Tomita, I., Oguni, I. and Konomoto, H. (1995) Effect of tea (Camilla sinesis L.) on lipid peroxidation in rat liver and kidney: a comparison of green tea feeding. Biol. Pharm. Bull. 18(7):1006–1008.

6

ANTIOXIDATIVE PROTEIN IN JAPANESE MUSHROOM

Shunro Kawakishi and Mitsuru Tanigawa

Department of Applied Biological Sciences
School of Agriculture
Nagoya University
Furo-cho, Chikusa-ku, 464-01 Nagoya, Japan

1. INTRODUCTION

Oxygen molecule incorporated to biological body is reduced to active oxygen species, as superoxide anion radical and etc, with metabolic systems in tissues and cell. These active oxygens are generally quenched by the action of the quenching and/or degrading enzymes as superoxide dismutase, however, the depression of their activities with endogeneous causes induce the oxidative damages in tissue and cell which may arise sometime to aging and diseases. The exogeneous factors as ascorbic acid, tocopherols and carotenoids also act for the scavenging or degrading of active oxygens and the suppression of these oxidative damages[1]. Moreover, natural antioxidants as polyphenols found in edible plants suppress the autoxidation of unsaturated fatty acids and inhibit the fomation of lipide peroxides and simultaneously they are also radical scavengers. The radical scavengers for hydroxyl radical in mushrooms have been investigated in details[2], but their active principles were not yet isolated and identified from mushroom extracts. We have investigated on the active principle as hydroxyl radical scavenger in Japanese mushuroom, Shiitake and isolated a kind of protein suppressing the oxidation with hydroxyl radical. This paper concerns with the isolation of active protein, its suppressive effects of the oxxidation reaction with hydroxyl radical and its another physiological activities.

2. HYDROXYL RADICAL SCAVENGING AND/OR QUENCHING ACTIVITIES OF THE EXTRACTS OF EDIBLE MUSHROOM

Cu(II)/H_2O_2 and Fe(II)-EDTA/H_2O_2 solutions were used as hydroxyl radical generating system, and the radical scavenging activities of the mushroom extracts were evaluated from the determination of malondialdehyde (MDA) from the oxidative degradation of 2-deoxyribose. From these results shown in Table I, the aqueous extracts of Shiitake,

Food and Free Radicals, edited by Hiramatsu *et al.*
Plenum Press, New York, 1997

Table I. Suppressive effects of mushroom extracts on oxidative degradation of 2-deoxyribose with hydroxyl radical

		Suppression of MDA formation(%)	
Mushroom	Extract added(mg)	Cu(II)/H2O2	Fe(II)-EDTA/H2O2
No addition	0	0	0
Shiitake(raw)	2	87.3	87.3
	0.2	46.1	66.7
Honshimeji	2	77.0	73.6
	0.2	29.8	54.0
Nameko	2	65.6	61.8
	0.2	2.8	39.8
Enokidake	2	64.2	88.9
	0.2	50.6	56.9
Awabidake	2	75.9	89.3
	0.2	5.3	60.7
Toramakidake	2	61.1	88.1
	0.2	48.4	54.3

To phosphate buffer solution(0.1M, pH 7.4) containing 2-deoxyribose (1 mM) and mushroom extract, $FeSO_4$ and $CuSO_4$(10μM), EDTA(10μM), H_2O_2(1 mM) were added and incubated at room temperature for 24h. MDA formed was determined by TBA method and the suppression(%) was calculated from their TBA values and controlled one.

Enokidake and Awabidake mushrooms exhibited the strong inhibitive activities to the MDA formation. For the research of active principle in Shiitake extract, the relation between its concentration and the inhibitive activity was investigated by the methods of 2-deoxyribose oxidation and benzoate hydroxylation, in which the later method was performed by the fluorescence determination of o-hydroxybenzoate with Em 407nm at Ex 305nm. The aqueous extract showed the inhibitive effects on these oxidation reactions dependently in its concentration of 2–200μg/mL(data not shown).

3. THE ISOLATION AND PURIFICATION OF ACTIVE PRINCIPLE IN SHIITAKE MUSHROOM

To isolate of the active principle exhibiting the hydroxyl radical scavenging and/or quenching activities, the aqueous extract of Shiitake was submitted to DEAE(Toyo Pearl 650M) column and eluted with water and followed by NaCl gradient (0 to 0.6%). The chromatography were performed on the determinations of suppressive activity by 2-deoxyribose oxidation method, protein by the UV absorption at 280nm and sugar moieties by phenol-sulfuric acid method (absorption at 490nm). As shown in Figure 1, the active fractions were determined as three peaks. The first fraction near Fr. 20 eluted with water was estimated as soluble carbohydrates from sugars determination. Other two active peaks

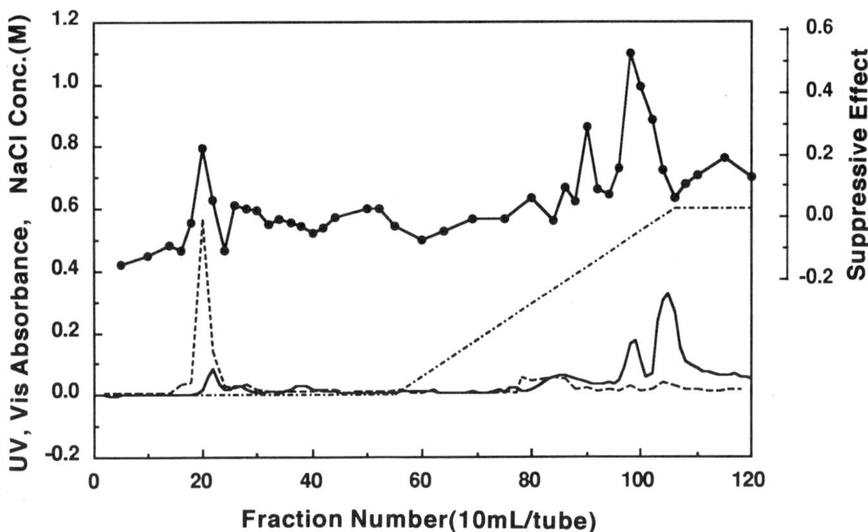

Figure 1. DEAE gel chromatography of fresh Shitake mushroom extract. Column: DEAE-TOYOPEARL 650 M. ●, suppressive effects (%) of oxidative degradation of 2-deoxyribose; - - - -, sugar detection with phenol-sulfuric acid method (E490 nm); ———, protein detection in UV absorption (E280 nm); — — — —, NaCl concentration (gradient method).

were the fractions eluted with NaCl gradient and the fractions 95–100 exhibited a most highest activity were assumed as a protein like substance from UV absorption at 280nm and its behavior on anion exchanger column. Then, we have tried the isolation and purification of this active protein from raw Shiitake. Raw Shiitake 2Kg was ground and extracted with phosphate buffer solution(0.01M, pH7.0) containing protease inhibitor and the extract was treated with 70% ammonium sulfate to precipitate the active protein. The precipitate dissolved in phosphate buffer solution (0.01M) was dialyzed and adsorbed to DEAE 650M (3x55cm) column. This was eluted with a similar manner in Figure. 1. The same active fraction with the fraction 95–100 was repeatedly purified with BUTYL TOYOPEARL 650S and TOYOPEARL HW-55S gel chromatography. The active protein obtained from above process gave a single peak on HPLC equipped with TSK-gel G3000SWXL column. Its molecular weight was determined as 107Kda from HPLC method and SDS-PAGE gave a single spot at 54Kda. Therefore, the active substance (named 107P) was a protein of 107Kda composed of two subunits of 54Kda.

4. SUPPRESSIVE EFFECTS OF ACTIVE PROTEIN 107P ON THE OXIDATION REACTION WITH ACTIVE OXYGEN

Suppressive effects of 107P on 2-deoxyribose with Fenton's reagent [Fe(II)-EDTA/H_2O_2] were compared with other active oxygen scavenging and quenching proteins, catalase, SOD, transferrin and human serum albumin(HSA). These proteins at several concentration levels were added to 2-deoxyribose oxidation system and their suppressive effects on oxidation were studied in detail. From Figure 2, 107P exhibited strong

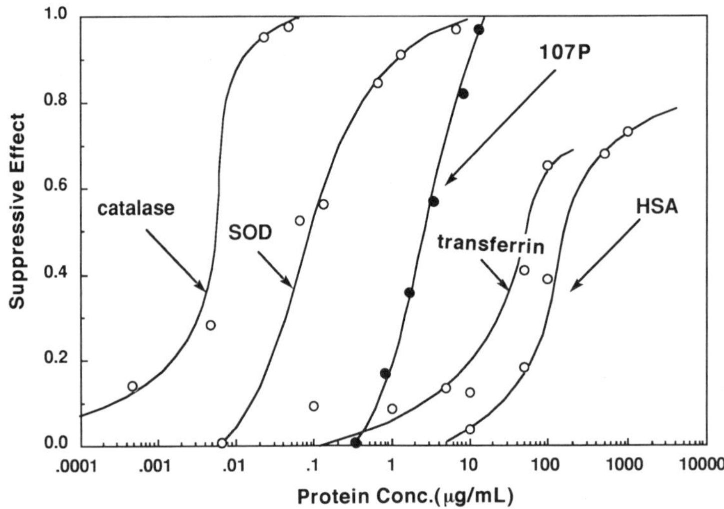

Figure 2. Suppressive effects of 107P and other active proteins on the oxidative degradation of 2-deoxyribose with hydroxyl radical.

suppressive effects at lower concentrations than transferrin and HSA which have transition metal-chelating abilities, but not so strong as same with catalase and SOD. The concentration of 107P for 50% suppression of 2-deoxyribose oxidation was 2.7µg/mL which corresponded to about 1/15 of the 50% suppressive concentration of Shiitake crude extract. The suppressive effects of 107P on hydroxylation of benzoate was also observed dependently with its concentration, however, its activity was a little lower than the case of 2-deoxyribose and its 50% suppressive concentration was 38µg/mL.

5. OTHER PHYSIOLOGICAL AND BIOCHEMICAL PROPERTIES OF 107P

When purified 107P was denatured by heating at 100°C for 10min and its activity was compared with native 107P, it was not inactivated perfectly and retained its activity about 50%. This means that 107P is a considerable thermostable protein. From its amino acid analysis, the contents of some amino acids as histidine, tyrosine, methionine, phenylalanine and proline which revealed all higher reactivities with hydroxyl radical, were lower compared with HSA. Therefore, this active protein would not be a scavenger of hydroxyl radical. Since the transition metal-containing oxidation systems were used for activity measurement, its activity may be related to the metal-binding or chelating properties. These properties between 107P and Fe(II) were researched in details, however, Fe-binding or chelating abilities of 107P could not be observed. Moreover, it was also investigated whether or not 107P have catalase, SOD and peroxidase-like activities. The result showed that 107P have not exhibited these enzyme activities. The another hydroxyl radical generating systems without transition metal ion, as hydrogen peroxide-UV and γ-irradiation systems, were applied to the oxidation of 2-deoxyribose, but 107P did not suppress its oxidative degradation. From these results, the suppressive effects of 107P on these oxidation were not due to scavenging or quencing action of hydroxyl radical and also metal chelating action, but it can therefore be presumed that 107P inhibits the formation of hydroxyl radical, especially transforms metal ion to oxidized inactive form. These estimation would be supported from fact that 107P did not also inhibits the oxidation of 2-deoxyribose by Fe(II)-EDTA/AsA system, using ascorbic acid, strong reductant, instead of hydrogen peroxide.

6. THE PROTECTIVE ACTION OF 107P TO OXIDATIVE INACTIVATION OF ENZYME

Metal catalyzed oxidation systems (MCO) containing transition metal ions have usually arisen an oxidative damage of proteins and enzyme inactivation in cell and tissue[3]. Then, the experiments on whether or not 107P suppresses the oxidative damage of enzyme and protects its activity were performed. Glutamine synthetase (GS) as model enzyme was used in this experiment. GS is easily oxidized and inactivated by the oxidative action of Fe(III)/AsA or Fe(II)/H_2O_2 and the oxidized GS tends to be hydolyzed with several proteases[4,5]. The oxidative damage of GS has been arisen at the metal binding site of enzyme molecule, and its histidine and arginine residues stereostructurally located near at the site were oxidized preferencially[6-8]. We have reconfirmed by the same method of Levine et al that Fe(II)(250μM)-H_2O_2(50μM) system inactivates GS most strongly. Then, 107P, SOD, transferin and HSA were added to this GS oxidation system at their several concentrations and their protective effects were investigated. From the results (Figure 3), 107P markedly suppressed the inactivation of GS dependently on its concentration. This means that 107P acts protectively for the oxidative damage of enzyme with hydroxyl radical generating system. Other proteins, SOD etc, also suppressed GS inactivation in lower level than that of 107P.

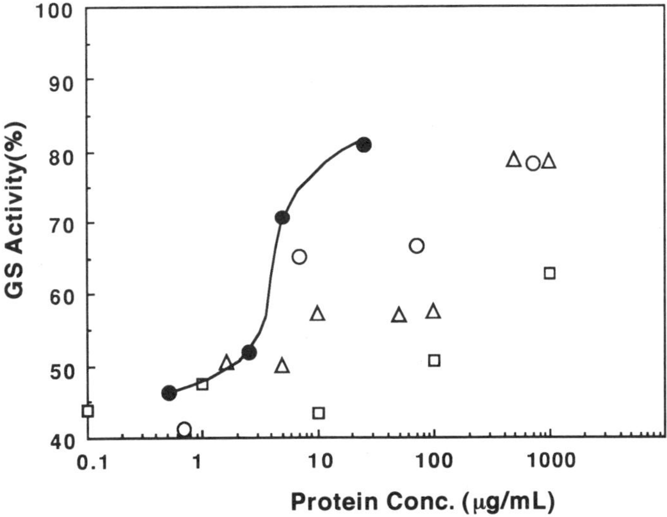

Figure 3. Suppressive effects of 107P and other active proteins on the inactivation of glutamine synthetase with hydroxyl radical. ●, 107P; △, human serum albumin (HSA); □, transferrin, ○, superoxide dismutase (SOD).

7. SUPPRESSIVE EFFECTS OF 107P ON THE OXIDATIVE CLEAVAGE OF DNA STRAND

It has been well known that DNA strand between 2-deoxyribose and phosphate ester are easily cleaved by the action of active oxygens and other radical species[9]. Especially, many works on the oxidative cleavage of ϕX174DNA have been presented[10–12]. Then, we have studied the inhibitive effect of 107P on the cleavage of ϕX174DNA with Fe(II)-EDTA/H_2O_2 system. This experiment was performed by the electrophoretic detection of relaxed and linear types formed from supercoiled type DNA with hydroxyl radical. By the preliminary experiments, ϕX174DNA(7.5ng) was oxidatively cleaved to relaxed type DNA with Fe(II)(10μM)-H_2O_2(100μM) system, but not to linear type. Then 107P, catalase, SOD, transferrin and HSA were added to this oxidation system at their several concentrations and incubated at 37°C for 15min. After incubation, the cleavage of this DNA was investigated by gel electrophoresis. The suppressive effect of 107P was shown in Figure 4. From this data, 107P perfectly inhibited the DNA cleavage at its concentration of 10μg/mL, however, other proteins exhibited strong quenching activities for active oxygens could not suppress the cleavage at same concentration with 107P(data not shown).

8. CONCLUSION

Active oxygen radicals and lipid peroxides have secondarily arisen the oxidation reaction in tissues and cells sometime inducing aging and diseases. Such oxidations in bio-

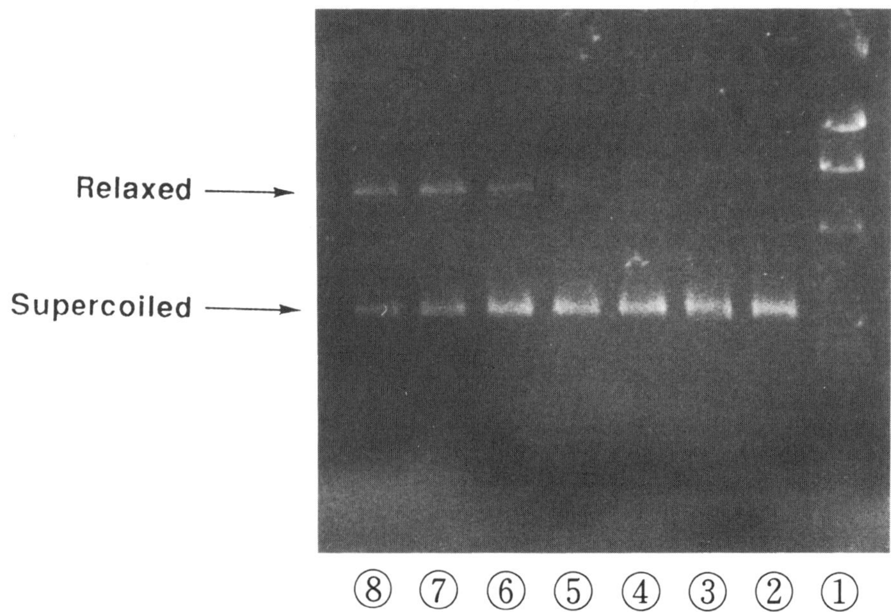

Figure 4. Suppressive effects of 107P on the strand cleavage of ϕX174 DNA with hydroxyl radical. 1, Marker, Hind III digest of λ phage as size standard; 2, control (ϕX174 DNA 7.5 ng); 3, DNA + 10 μM Fe(II); 4–8, DNA 7.5 ng + Fe(II) 10 μM + H_2O_2 100 μM + 107P: (4) 50 μg/mL, (5) 10 μg/mL, (6) 5 μg/mL, (7) 1 μg/mL, (8) 0 μg/mL.

logical cells are usually controlled with active oxygens quenching and scavenging enzyme systems and also tocopherol, ascorbic acid and carotens as food originating factors. Antioxidants as the later, said to be antioxidative vitamins, and polyphenols distributed widely in plant origins are daily taken with foods, especially vegetable, fruits and grains. Among them, some one is incorporated to several tissues and may be functioned its biological activity as tocopherol. From the facts that 107P inhibited the oxidation reaction with hydroxyl radical generationg system, it might be also an antioxidative protein. However, the protein is neither a scavenging factor of hydroxyl radical as SOD and peroxidase, nor a metal chelator. It may be related to oxidation or reduction of transition metal ions in active oxygen generationg system.

SUMMARY

We have isolated an active protein suppressing the oxidation with hydroxyl raical from Japanese mushroom, Shiitake. The protein named 107P was 107KDa as molecular weight and composed of two subunits of 54KDa. 107P inhibited 50% of the oxidation of 2-deoxyribose with hydroxyl radical at its concentration of 2.7µg/mL. 107P did not exhibites catalase, SOD and peroxidase-like activities and also Fe-binding of chelating abilities. However, 107P have shown the marked protective action on inhibition of glutamine synthetase by metal-catalyzed oxidation systems and also suppressive effects on strand cleavages of ϕX174DNA with hydroxyl radical. From these results, it may be that the suppressive actions of 107P are related on the inhibition of hydroxyl radical generation system.

REFERENCES

1. Niki, E. (1988) Ascorbic acid, Glutathione. Protein, Nucleic Acid and Enzyme. 33:2973–2986.
2. Hiramatsu, M., Edamatsu, R., Kohno, M. and Mori, A. (1989) Active oxygen quenching action of edible mushroom. In: ESR and Free Radical, pp155–159. Nihon Igaku Kan, Tokyo.
3. Stadtman, E.R. (1990) Metal ion catalyzed oxidation of proteins.: Biochemical mechanism and biological consequences. Free Rad. Biol. Med. 9:315–325.
4. Levine, R.L. (1983) Oxidative modification of glutamine synthetase. I. Inactivation is due to loss of one histidine residue. J. Bipol. Chem. 258:11823–11827.
5. Levine, R.L. (1983) Oxidative modification of glutamine synthetase. II. Chracterization of the ascorbate model system. J. Biol. Chem. 258:11828–11833.
6. Farber, J.M. and Levine, R.M. (1986) Sequence of a peptide susceptable to mixed- function oxidation. J. Biol. Chem. 261:4574–4578.
7. Rivett, A.J. and Levine, R.L. (1990) Metal-catalyzed oxidation of Echerichia coli glutamine synthetase: Correlation of structure and functional changes. Arch. Biochem. Biophys. 278:26–34.
8. Climent, I., and Levine, R.L. (1991) Oxidation of the active site of glutamine synthetase conversion of arginine-344 to g-glutamyl semialdehyde. Arch. Biochem. Biophys. 289:371–375.
9. Nagano, T. and Hirobe, M. (1988) Degradation of nucleic acids by active oxygen species. Protein Nucleic Acid and Enzyme. 33:3094–3101.
10. Rozenberg-Arska, M., Asbek, S. van, Martens, T.F.J. and Verfoef, J. (1985) Damage to chromosomal and plasmid DNA by toxic oxygen species. J.Gen. Microbiol. 131:3325–3330.
11. Lafleur, M.V.M., Nienwint, A.W.M., Aubry, J.M., Kortbeek, H., Arwert, F. and Joenje, H. (1987) DNA damage by chemically generated singlet oxygen. Free Radical Res. Commun. 2:343–350.
12. Sagripanti, J.L. and Kraemer, K.H. (1989) Site specific oxidative DNA damage at polyguanosines produced by copper plus hydrogen peroxides. J. Biol. Chem. 264:1729–1734.

7

ANTIOXIDANT ACTION OF GINKGO BILOBA EXTRACT (EGB 761)

Lester Packer[*]

251 Life Sciences Addition
Department of Molecular and Cell Biology
University of California
Berkeley, California

1. INTRODUCTION

Extracts from the leaves of *Ginkgo biloba* trees have been used therapeutically for centuries in traditional Chinese medicine, and in modern Chinese pharmacopoeias both the leaves and fruits are recommended for treating problems of heart and lungs. EGb 761 is a dry, powdered extract prepared from *Ginkgo biloba* leaves. It is a standardized mixture of several different chemical constituents; its two major classes of compounds are flavonoid glycosides and terpenoids (Drieu 1986; DeFeudis 1991). The flavonoid fraction is mainly composed of three flavonols: quercetin, kaempferol, and isorhamnetin, which are linked to a sugar(DeFeudis 1991). The terpenoid fraction is composed of ginkgolides and bilobalides(Drieu 1988). It also contains some organic acids, which help make it water-soluble(Drieu 1986).

EGb 761 has been shown to have a superoxide dismutase-like activity and hydroxyl radical scavenging activity (Pincemail and Deby 1986; Pincemail, Dupuis et al. 1989). Furthermore, it has been recently reported that EGb 761 has the property of inactivating oxo-ferryl radical species that are more efficient oxidative agents than classical hydroxyl radicals (Deby, Deby-Dupont et al. 1993). In the past several years we have investigated the effectiveness of EGb 761 against a variety of radical species, especially nitric oxide (NO), and in the treatment of ischemia-reperfusion injury and the prevention of LDL oxidation. We have recently also examined the effects of this versatile mixture of antioxidants in influencing gene expression through its influence on transcription factors.

---

* Address for correspondence: Lester Packer, 251 Life Sciences Addition, Department of Molecular and Cell Biology, University of California, Berkeley, CA 947203-200, Phone: 5106-42-1872, Fax: 510-642-8313.

Food and Free Radicals, edited by Hiramatsu *et al.*
Plenum Press, New York, 1997

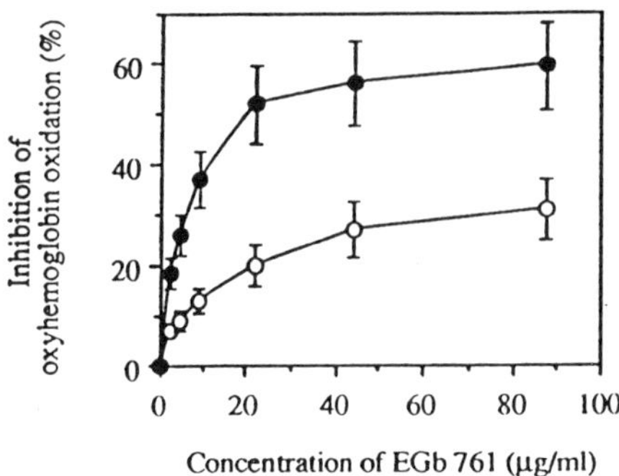

Figure 1. Effect of EGb 761 on hemoglobin oxidation by nitric oxide generated from the reaction of Complex I of catalase with hydroxylamine. Solutions of hemoglobin and a concentration of 6 uM (closed circles) and 25 uM (open circles) were exposed, in the presence of different concentrations of EGb 761 at 37 °C, to nitric oxide generated by the enzymatic system. Each point represents the mean ± S.D. of 4 experiments.

2. EFFECTS ON NITRIC OXIDE

Among the molecular mechanisms that may underly the vascular effects of EGb 761, is has been hypothesized to enhance the cellular action of endothelium-derived relaxing factor (EDRF) (DeFeudis 1991), which is known to relax the muscular cells of blood vessels and to inhibit platelet aggregation by the activation of guanylate cyclase and the consequent increase of the concentration of cellular cGMP (Mocada, Palmer et al. 1991). The chemical nature of EDRF is unclear, but it is thought to be nitric oxide (NO). In light of the multicomponent chemical nature of EGb 761 and the properties of the extract to scavenge efficiently not only superoxide but also other oxygen radicals such as hydroxyl radical and peroxyl radical, we investigated the possible NO-scavenging properties of EGb 761.

Oxyhemoglobin is a well known scavenger of nitric oxide, which oxidizes the protein to the methemoglobin form with concomitant production of nitrate (Doyle and Hoekstra 1981). Scavengers of nitric oxide should compete with the hemoglobin for nitric oxide and affect the rate of hemoglobin oxidation depending on their concentration. EGb 761 inhibited the oxidation of hemoglobin to methemoglobin in a dose-dependent manner **(Fig 1)**, and the inhibition was dependent on the concentration of hemoglobin **(Fig 2)**. The nitric oxide in these experiments was produced by reaction of hydroxylamine with complex I of catalase under essentially the conditions reported by Murphy and Sies (Murphy and Sies 1991).

The inhibition by EGb 761 of oxidation of hemoglobin by nitric oxide indicates that the EGb 761 was effectively scavenging the NO. To confirm this, we used a different system for generation and detection of NO. Nitric oxide was generated from sodium nitroprusside, which decomposes in aqueous solution at physiological pH to produce NO

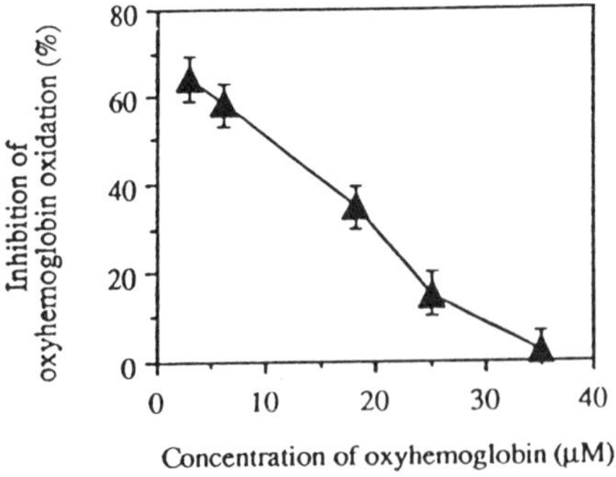

Figure 2. Inhibition of oxyhemoglobin oxidation by 80 uM EGb 761 at various hemoglobin concentrations. Measurements were at 37 °C and each point represents the mean ± S.D. of 4 experiments.

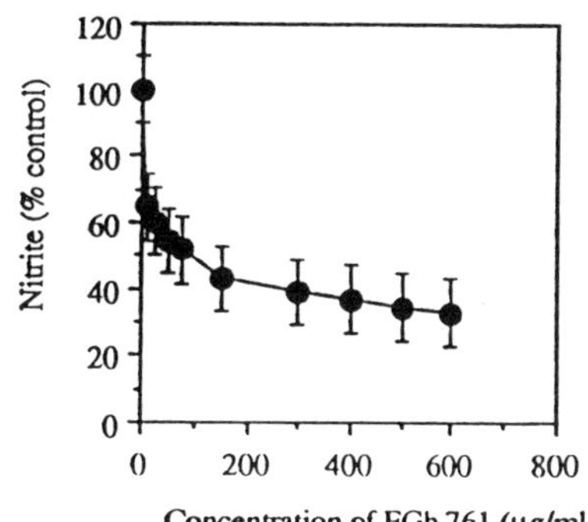

Figure 3. Effect of EGb 761 on the amount of nitrite generated during incubation with 5 mM sodium nitroprusside. Incubation time 150 min., temperature 20 °C. Each point represents the mean ±S.D. of 3 experiments.

(Feelisch and Noack 1987), and NO was detected through formation of nitrite produced by the reaction of NO with oxygen, using the Griess reagent (Green, Wagner et al. 1982). Again, in this system, EGb 761 decreased NO concentration in a dose-dependent manner **(Fig 3)**; the maximum effect occurred at 300 ug/ml EGb 761, and the half-maximum effect occurred at 20 ug/ml.

Nitric oxide plays both a regulatory role (e.g. as a vasodilator) and a cytotoxic role (e.g., in the macrophage). The NO-scavenging properties of EGb 761 we observed expand the role of EGb 761 as an antioxidant agent, and the fact that EGb 761 affects a substance that acts on the vasculature may help explain its effectiveness in a variety of conditions in which blood flow is affected. However, although an endothelium-dependent vasorelaxant effect of EGb 761 has been observed (DeFeudis 1991), factors other than EDRF/NO could be involved in the beneficial vascular effects of EGb 761. For example, it has been reported that EGb 761 stimulated the release of prostacyclins from the endothelium of rat aortic preparations (Chaterjee 1985). In cultured bovine endothelium cells, NO has been shown to inhibit the release of these unstable arachidonate metabolites which are potent vasodilators and inhibitors of platelet aggregation ; thus, by scavenging NO EGb 761 could potentiate the biological effect of prostacyclins.

3. EFFECTS IN ISCHEMIA-REPERFUSION

EGb 761, with its proven antioxidant properties, is a logical candidate to protect against ischemia-reperfusion injury of the heart. During ischemia-reperfusion of the heart, oxygen radicals are thought to play an important role in the genesis of tissue injury(McCord 1985), and there is much evidence that reactive oxygen species such as the superoxide anion, hydrogen peroxide, and hydroxyl radical are involved in reperfusion injury of the post-ischemic heart(Brown, Terada et al. 1988; Zweier, Kuppusamy et al. 1989; Terada, Rubinstein et al. 1991). Free radical scavengers have been demonstrated to enhance functional recovery of the heart (Janero and Burghardt 1989; Packer, Valenza et al. 1991). Hence, we undertook studies of the effects of EGb 761 on ischemia-reperfusion in the heart. We investigated the influence of EGb 761 on cardiac ischemia-reperfusion injury such as lower functional recovery and enzyme leakage during reperfusion, and we also investigated its effect on endogenous antioxidants such as ascorbate, in order to understand the relationship between its cardioprotective effects and its antioxidant properties.

Male Sprague-Dawley rats were anesthetized and the heart was excised and the aorta was immediately attached to a perfusion apparatus using the Langendorff method

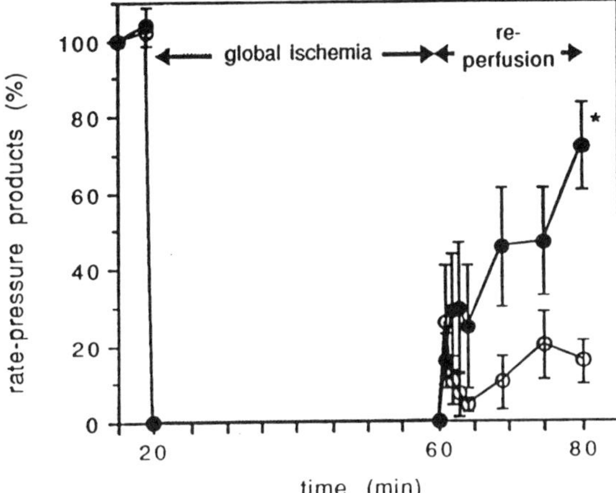

Figure 4. Effect of EGb 761 on cardiac mechanical recovery during reperfusion. Rate-pressure products were calculated as (heart rate x left ventricular pressure). Functional recoveries were expressed as a percentage compared with function of the preischemic period. Each absolute value of control (open circles) and EGb 761 (closed circles) treated hearts during preischemic period was 205.2 ± 32.8 and 226.6 ± 34.0 (102 x mm Hg x beat/min), respectively. Each value was expressed as mean ± SE. Numbers of each group are greater than 4. *$p < 0.01$ compared with the absence of EGb 761.

(Langendorff 1895). Another cannula was inserted into the left ventricle via the left atrium and connected to a pressure transducer. Perfusate was a modified Krebs buffer with and without 200 mg/l EGb 761. An initial 20 min equilibration period was followed by an additional 40 min of "no-flow" normothermic global ischemia and 20 min of reperfusion. Control hearts were subjected to 60 min perfusion (no ischemia) with and without 200 mg/l EGb 761.

Cardiac mechanical recovery after ischemia was greatly improved by treatment with EGb 761 **(Fig 4)**. Control hearts exhibited less than 20% recovery of mechanical function in the post-ischemic period, whereas hearts exposed to perfusate containing EGb 761 recovered to 72% of initial mechanical rate-pressure values. The time course of enzyme (lactate dehydrogenase, LDH) leakage during ischemia and reperfusion was also monitored, as a marker of cell membrane integrity. Hearts treated with EGb 761 showed essentially no enzyme leakage, while values for control heart increased almost 100%. Myocardial tissue total ascorbate content and ascorbate (reduced form) were significantly decreased after 20 min of reperfusion following 40 min of global ischemia without EGb 761 **(Fig. 5)**. Moreover, the percentage of dehydroascorbate (oxidized form) was significantly increased. These changes in ascorbate suggest that ascorbate acted as an antioxidant during this period of ischemia-reperfusion. Interestingly, this decrease in ascorbate and increase in dehydroascorbate content were both suppressed by 200mg/l EGb 761. In addition, there was no significant change in the ascorbate or dehydroascorbate level after 60 min perfusion performed with and without EGb 761.

The low mechanical recovery and enzyme leakage of hearts subjected to ischemia-reperfusion, provide evidence of damage. The significant improvement in mechanical recovery and decrease in LDH leakage in hearts perfused with 200 mg/l EGb 761 indicate that this concentration of EGb 761 protects the heart against the ischemia-reperfusion damage produced by the experimental conditions.

The results also show that EGb 761 prevents the leakage and oxidation of ascorbate, and suggest two possibilities for a protective mechanism against cardiac ischemia-reperfusion injury. One is that EGb 761 can work as a primary preventive antioxidant in place of ascorbate. The other possibility is that EGb 761 reduced the oxidized ascorbate after ascorbate worked as the primary antioxidant. To test the former possibility, we tested the ability of EGb 761 to react with the stable free radical diphenylpicrylhydrazyl (DPPH) in

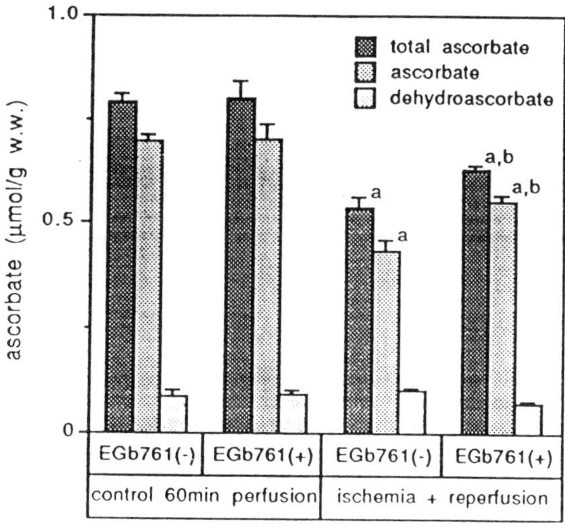

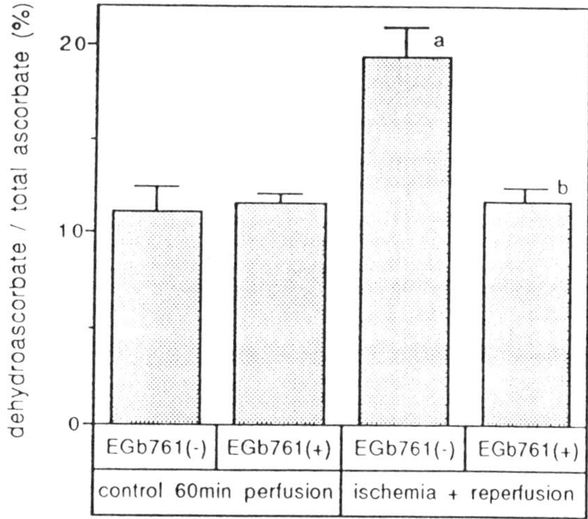

Figure 5. Influence of EGb 761 on myocardial tissue ascorbate content. (A) Total ascorbate, ascorbate (reduced form), dehydroascorbate (oxidized form). (B) Percentage of dehydroascorbate was compared to total ascorbate. Each value was expressed as mean ±SE. Numbers of each group are greater than 4. *a $p < 0.05$ compared with 60 min control perfusion of each group. *b $p < 0.05$ compared with ischemia-reperfusion group perfused by control buffer.

a dose-dependent manner, using electron spin resonance (ESR) to detect the radical (**Fig 6**). EGb 761 quenched DPPH radicals, and the maximum stable quenching effect was observed at a concentration of 200 mg/l, precisely that which was used for cardiac perfusion experiments. Hence, the protective effect of EGb 761 against cardiac ischemia-reperfusion injury might depend on its free radical-quenching properties. This view is supported by recent work by Shen and Zhou (Shen and Zhou 1995), in which treatment with EGb 761 decreased lipid peroxidation during ischemia-reperfusion in rabbit hearts.

4. PREVENTION OF OXIDATION OF LOW DENSITY LIPOPROTEIN

Oxidative damage to biological lipids is involved in the pathophysiology of cerebral and peripheral vascular diseases, for which EGb 761 is used as a treatment (DeFeudis

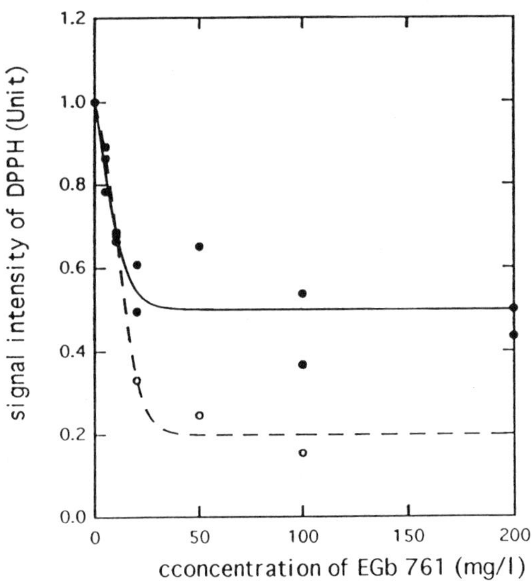

Figure 6. Dose dependency of quenching of DPPH by EGb 761. ESR spectra were recorded at room temperature. Arbitrary intensity of DPPH signal was plotted against the concentration of EGb 761. Closed circles represent EGb 761 in Krebs-bicarbonate buffer. Open circles represent EGb 761 in ethanol.

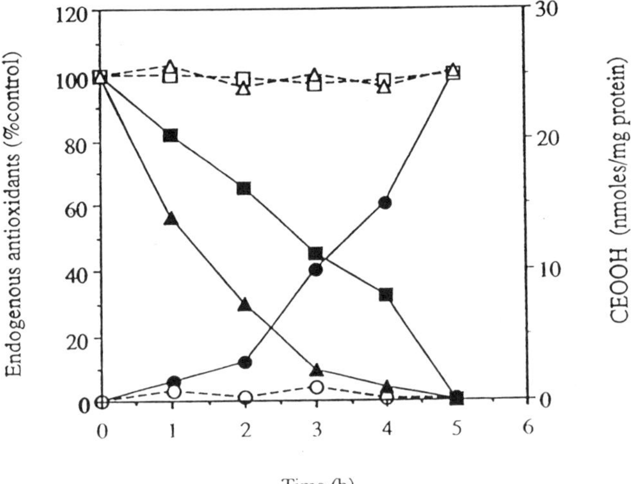

Figure 7. Effect of EGb 761 on oxidative modification of human LDL induced by AAPH. Human LDL (1 mg protein/ml) was incubated in PBS at 37 °C with AAPH (5mM) in the absence (solid lines, closed symbols) and in the presence (broken lines, open symbols) of EGb 761 (100 ug/ml). Cholesterol linoleate ester hydroperoxide (circles); α-tocopherol, initial concentration: 2.5 nmol/mg protein, (triangles); β-carotene, initial concentration: 0.27 nmol/mg protein (squares). Endogenous antioxidants are expressed as a percentage of the level measured in control samples incubated in the absence of azo-initiator. In control samples formation of cholesterol linoleate ester hydroperoxide, depletion of α-tocopherol or β-carotene were not observed. Means ± SD obtained from triplicate assays within a representative experiment out of three are reported; in some cases error bars are smaller than the symbols.

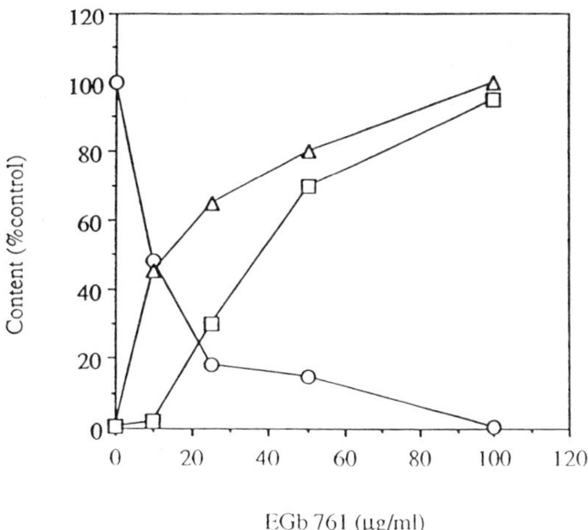

Figure 8. Concentration effects of EGb 761 on oxidative modification of human LDL induced by AAPH. Human LDL (1 mg protein/ml) was incubated for 5 hours in PBS at 37 °C with AAPH (5 mM) in the presence of various concentrations of EGb 761. Cholesterol linoleate ester hydroperoxide (circles) is expressed as a percentage of the level measured in control samples incubated in the absence of EGb 761 (i.e. 65, 25, 93 nmol/mg protein for LDL isolated from three different donors). The content of α-tocopherol (initial concentration for LDL isolated from three different donors: 6.5, 2.5, 10 nmoles/mg protein) (triangles), and of β-carotene (initial concentration for LDL isolated from three different donors: 0.18, 0.27, 0.16 nmol/mg protein) (squares) are expressed as percentage of the levels measured in control samples incubated in the absence of azo-initiator. Means ± SD obtained by using the three different LDL preparations; in some cases error bars are smaller than the symbols.

1991; Kleijnen and Knipschild 1991). Such damage is also involved, however, in a variety of other chronic diseases such as cancer, cataract, and atherosclerosis (Halliwell and Gutteridge 1989). Thus, to evaluate the peroxyl scavenging activity of EGb 761 in a physiologically relevant model of damage to lipid systems, we analyzed the effect of the extract on oxidative damage to human low density lipoprotein (LDL). Lipoproteins were exposed to azo-initiators and the effects of the extract analyzed in terms of: 1) accumulation of cholesterol linoleate ester hydroperoxides; 2) depletion of α-tocopherol and β-carotene; and 3) changes in intrinsic tryptophan fluorescence.

PBS suspensions of LDL at a final concentration of 1 mg protein/ml were used in all experiments. LDL suspensions, in the presence or absence of EGb 761, were preincubated at 37o C for 10 min. AAPH and AMVN (5 mM or 1mM final concentration, respectively) was added and incubated for 5 hours with agitation in a water bath under air. At timed intervals, aliquots of the incubation mixture were collected and analyzed for cholesterol linoleate ester hydroperoxides, α-tocopherol, and β-carotene by HPLC.

Treatment of human LDL with peroxyl radicals generated from AAPH caused a progressive accumulation of cholesterol linoleate ester hydroperoxide and a concurrent loss of α–tocopherol and β–carotene **(Fig. 7)**. Although different LDL samples had different concentrations of endogenous antioxidants and showed different rates of oxidation, they all showed similar trends of oxidative modification: the rate of depletion of α–tocopherol was higher than that of β–carotene. In all LDL preparations, EGb 761 (100 ug/ml) completely protected LDL from oxidative modification: no decreases in α–tocopherol or β–carotene were observed, nor was there an accumulation of cholesterol ester hydroperoxides. The protective effect of EBb 761 was concentration-dependent **(Fig. 8)**. EGb 761 at a concentration of 10 ug/ml efficiently protected LDL from the accumulation

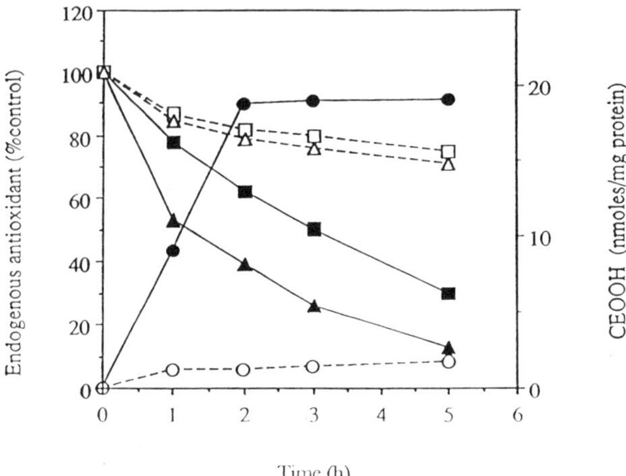

Figure 9. Effect of EGb 761 on oxidative modification of human LDL induced by AMVN. Human LDL (1 mg protein/ml) was incubated in PBS at 37 °C with AMVN (1 mM) in the absence (solid lines, closed symbols) and in the presence (broken lines, open symbols) of EGb 761 (100 ug/ml). Cholesterol linoleate ester hydroperoxide (circles); α-tocopherol, initial concentrations: 3.9 nmoles/mg protein (triangles); β-carotene, initial concentrations: 0.21 nmol/mg protein (squares). Endogenous antioxidants are expressed as percentage of the levels measured in control samples incubated in the absence of azo-initiator. In control samples, the formation of cholesterol linoleate ester hydroperoxide and the depletion of α-tocopherol or β-carotene were not observed. Means ± SD obtained from triplicate assays within a representative experiment out of three are reported; in some cases error bars are smaller than the symbols.

of cholesterol ester hydroperoxides and from the depletion of α–tocopherol, but the extract at this concentration did not protect LDL from β–carotene depletion.

EGb 761 was also able to protect LDL against oxidative damage induced by the lipid-soluble peroxyl radical generating compound AMVN. Although different LDL samples had different concentrations of endogenous antioxidant and showed different rates of oxidation, they all showed similar trends of oxidative modification. The exposure of suspensions of LDL to AMVN (1mM) (resulted in an accumulation of cholesterol linoleate ester hydroperoxides, and a depletion of α–tocopherol and β–carotene **(Fig. 9)**. The rate of tocopherol depletion was higher than that of β–carotene. In the presence of EGb 761 (100 ug/ml) the formation of cholesterol linoleate ester hydroperoxide was almost completely inhibited. The extract also protected LDL against depletion of α–tocopherol and β–carotene **(Fig. 9)**.

AAPH generates peroxyl radicals in aqueous environments, while AMVN generates peroxyl radicals in lipid environments. The rates of oxidation of the different components of LDL are consistent with an initial formation of radicals mostly outside or at the surface of LDL. After depletion of most of the tocopherol, peroxidation takes place inside the lipoprotein as a consequence of propagation of lipid peroxidation chain reactions. The ability of EGb 761 to protect α–tocopherol appeared greater than its ability to protect β–carotene; this was especially apparent at low concentrations of EGb 761 **(Fig 8)**. This could be due to a lower antioxidant activity of the molecules of EGb 761 in the LDL against peroxyl radical generated in the lipid domain, as well as to an unfavorable location of the antioxidant molecules of EGb 761 with respect to β–carotene.

EGb 761 contains both flavonoid glycosides and terpenoids. Flavonoids have been reported to be effective scavengers of superoxide (Yuting, Rongliang et al. 1990), hydroxyl radicals(Husain, Cillard et al. 1987), and inhibitors of lipid peroxidation (Afanas'ev, Dorozhko et al. 1989). Our results do not allow us to distinguish which component of EGb 761 is responsible for its scavenging properties against peroxyl radicals. As EGb 761 is a mixture of different chemical constituents, its scavenging activity could be due to a particular component as well as to the interactions of different antioxidant molecules.

The ability of EGb 761 to protect LDL against oxidative modification suggests a potential therapeutic use in the pathogenesis of atherosclerosis, which has been reported to be partly due to oxygen radical-induced modification of lipid or protein components of human LDL (Esterbauer, Gebicki et al. 1992). Accumulation of foam cells is characteristic of atherosclerotic lesions (Goldstein and Brown 1977; Scaffner, Taylor et al. 1980), and foam cells may be the result of accumulation of cholesterol, after excessive uptake of oxidized LDL by artery wall macrophages and smooth muscle cells (Goldstein, Ho et al. 1979). Formation of oxidized LDL as a result of free radical-initiated reactions in the body has also been hypothesized (Morel, Hessler et al. 1983).

REFERENCES

Afanas'ev, I. B., A. I. Dorozhko, et al. (1989). "Chelating and free radcial scavenging mechanism of inhibitory action of rutin and quercetin in lipid peroxidation." *Biochem. Pharmacol.* **38**: 1763–1769.

Brown, J. M., L. S. Terada, et al. (1988). "Hydrogen peroxide mediates reperfusion injury in isolated rat heart." *Mol. Cell. Biochem.* **84**: 173–175.

Chaterjee, S. S. (1985). . *Effect of Ginkgo biloba on organic cerebral impairment.* A. Agnoli, J. R. Rapin, V. Scapagnini and W. V. Weitbrecht. London, John Libbey.

Deby, C., G. Deby-Dupont, et al. (1993). "Efficiency of Ginkgo biloba extract (EGb 761) in neutralizing ferryl ion-induced peroxidations: Therapeutic implications." *Advances in Ginkgo biloba Extract Research* **2**: 13–26.

DeFeudis, F. V. (1991). *Ginkgo biloba extract (EGb 761): Pharmacological activities and clinical applications.* Paris, Elsevier.

Doyle, M. P. and J. W. Hoekstra (1981). "Oxidation of nitrogen oxides by bound dioxygen in hemoproteins." *J. Inorg. Biochem.* **14**: 351–358.

Drieu, K. (1986). "Preparation et definition de l'extrait de *Ginkgo biloba.*" *Presse. Med.* **15**: 1455–1457.

Drieu, K. (1988). . *Rokan, Ginkgo biloba: Recent Results in Pharmacology and Clinic.* E. W. Funfgeld. Berlin, Springer-Verlag: 32.

Esterbauer, H., J. Gebicki, et al. (1992). "The role of lipid peroxidation and antioxidants in oxidative modifications of LDL." *Free Rad. Biol. Med.* **13**: 341–390.

Feelisch, M. and E. A. Noack (1987). "Correlation between nitric oxide formation during degradation of organic nitrates and activation of guanylate cyclase." *Eur. J. Pharmacol.* **139**: 19–30.

Goldstein, J. L. and M. S. Brown (1977). "The low density lipoprotein pathway and its relation to atherosclerosis." *Ann. Rev. Biochem.* **46**: 897–930.

Goldstein, J. L., Y. K. Ho, et al. (1979). "Binding site on macrophages that mediates uptake and degradation of acetylated low density lipoprotein, producing massive cholesterol deposition." *Proc. Natl. Acad. Sci. USA* **76**: 33–37.

Green, L. C., D. A. Wagner, et al. (1982). "Analysis of nitrate, nitrite, and [15N]nitrate in biological fluids." *Anal. Biochem.* **126**: 131–138.

Halliwell, B. and J. M. C. Gutteridge (1989). *Free radicals in biology and medicine, second edition.* Oxford, England, Clarendon Press.

Husain, S. R., J. Cillard, et al. (1987). "Hydroxyl radical scavenging activity of flavanoids." *Phytochemistry* **26**: 2489–2491.

Janero, D. R. and B. Burghardt (1989). "Oxidative injury to myocardial membrane: direct modulation by endogenous alpha-tocopherol." *J. Mol. Cell. Cardiol.* **21**: 1111–1124.

Kleijnen, J. and P. Knipschild (1991). "Ginkgo biloba." *Lancet* **340**: 1136–1139.

Langendorff, O. (1895). "Untersuchungen am uberlebenden Saugertierherzen." *Pflugers Arch* **61**: 291–332.

McCord, J. M. (1985). "Oxygen-derived free radicals in postischemic tissue injury." *New Engl. J. Med.* **312**: 159–163.

Mocada, S., R. M. J. Palmer, et al. (1991). "Nitric oxide: physiology, pathophysiology, and pharmacology." *Pharmacol. Rev.* **43**: 109–142.

Morel, D. W., J. R. Hessler, et al. (1983). "Low density lipoprotein cytotoxicity induced by free radical peroxidation of lipids." *J. Lipid Res.* **24**: 1070–1076.

Murphy, M. E. and H. Sies (1991). "Reversible conversion of nitroxyl anion to nitric oxide by superoxide dismutase." *Proc. Natl. Acad. Sci.* **88**: 10860–10864.

Packer, L., M. Valenza, et al. (1991). "Free radical scavenging is involved in the protective effect of L-propionyl carnitine against ischemia-repurfuions injury of the heart." *Arch. Biochem. Biophys.* **288**: 533–537.

Pincemail, J. and C. Deby (1986). "Proprietes antiradicalaires de l'estrait de Ginkgo biloba." *Presse Med.* **15**: 1475–1479.

Pincemail, J., M. Dupuis, et al. (1989). "Superoxide anion scavenging effect and superoxide dismutase activity of Ginkgo biloba extract." *Experientia* **45**: 708–712.

Scaffner, T., K. Taylor, et al. (1980). "Arterial foam cells with distinctive immunomorpholitic and histochemical features of macrophages." *Am. J. Pathol.* **100**: 57–80.

Shen, J.-G. and D.-Y. Zhou (1995). "Efficiency of Ginkgo biloba extract (EGb 761) in antioxidant protection against myocardial ischemia and reperfusion injury." *Biochem. Mol. Biol. Int.* **35**: 125–134.

Terada, L. S., J. D. Rubinstein, et al. (1991). "Existence and participation of the xanthine oxidase in reperfusion injury of ischemic rabbit myocardium." *Am. J. Physiol.* **260**: H805-H810.

Yuting, C., Z. Rongliang, et al. (1990). "Flavanoids as superoxide scavengers and antioxidants." *Free Rad. Biol. Med.* **9**: 19–21.

Zweier, J. L., P. Kuppusamy, et al. (1989). "Measurement and characterization of postischemic free radical generation in the isolated perfused heart." *J. Biol. Chem.* **264**: 18890–18895.

8

ANTIOXIDATIVE COMPOUNDS FROM MARINE ORGANISMS

Kanzo Sakata

Research Laboratory of Marine Biological Science
Faculty of Agriculture
Shizuoka University
1944-2 Shinkawa, Mochimune
Shizuoka 421-01, Japan

1. INTRODUCTION

Radicals are shown in Chapter I to take part in lipid peroxidation which causes aging in organisms and cancer promotion as well as food deterioration. Antioxidants act as radical scavengers. Many efforts have been made for new antioxidants from terrestrial plants[1-3] and a few from marine organisms. Table I shows a brief review on chemical studies for antioxidative compounds from marine organisms.

Marine algae are well known that their quality is not changed during storage in spite of their high content of highly unsaturated fatty acids. Several studies have been conducted to isolate antioxidants from algae using the conventional antioxidant test[4-10]. Four kinds of bromophenols were isolated from a red alga[6]. Phlorotannins from the brown alga *Sargassum thunbergii*[7]; pheophytin from a green alga[8]. Phospholipid fractions containing phosphatidylcholine etc. are claimed to be active constituents in many cases, but the real active constituents have not been clear-cut. The presence of synergists for tocopherols in some algal extracts was also reported[10]. But nothing is clear chemically. Only a few antioxidants have been isolated from marine animals[11, 12]; Yamazaki, 1975 #46]. Knowing these back-ground, we attempted to find new antioxidative compounds from marine organisms.

2. MATERIALS AND METHODS

2.1. Extraction and Fractionation of Fish and Bivalves Viscera

Details refer to the previous paper[13, 14]

Food and Free Radicals, edited by Hiramatsu *et al.*
Plenum Press, New York, 1997

Table I. Chemical studies from antioxidants from marine organisms

Marine bacteria:
  Agrobacterium auranticam: Astaxanthin and 4-ketozeaxanthin[25]
Algae:
  Red algae:
    Porphyra sp. ("Nori"): Phospholipid fraction (Phosphatidylcholine)
    Polysiphonia urceolata ("Shojokenori"): 5-Bromo-3,4-dihydroxybenzaldehyde and a few related
      compounds[6]
  Brown algae:
    Eisenia bicyclis ("Arame"): Phosphatidylcholine and phosphatidylethanolamine[5]
    Undaria pinnatifida ("Wakame"): Unidentified antioxidants and synergists ( )[4]
    Laminaria sp. ("Konbu"): Unidentified antioxidants and synergists with α-tocopherol[10]
    Hizikia fusiformis ("Hijiki"): Unidentified antioxidants[5]
    Sargassum thunbergii ("Umitoranoo"): Phlorotannins[7]
    Cystoseria stricta: Phenolic meroditerpenoids[26]
  Green algae:
    Enteromorpha sp. ("Aonori"): Pheophytin a[4,8]
  Microalgae:
    Haematococcus pluvialis: Astaxanthin (from cyst cells)[27]
Marine animals:
  Krills: α-Tocopherol and phospholipid fraction (phosphatidylcholine)[12]; astaxanthin
  Prawns and Squids: Tocopherols and trimethylamine oxide[11]

2.2. A Screening Method for Antioxidants Using Paper Disks

On a paper disk (ϕ8 x 1.5 mm) for an conventional antimicrobial test, 2.0 mg of methyl linolate [20 μl of ethanolic solution (0.1 g/ml)] and 20 μl of each sample solution (equivalent to 5 mg of the extract) were absorbed as evenly as possible after mixing them well in a microsyringe. The disk was dried in a desiccator and incubated at 37°C in the dark. Each disk was subjected to a POV determination method modified the conventional American Oil Chemists' Society Method.

2.3. Mutagenicity

The paper disk prepared in the same way as mentioned above (sample equivalent to 3 mg of each extract was applied) was subjected to the rec-assay method developed by Kada et al.[15] to measure mutagenicity.

2.4. A Modified Ferric Thiocyanate Method

Ferric thiocyanate method by Fukuda *et al.*[16] was modified as follows. Each sample dissolved in appropriate solvent (*ca.* 30 μL) was absorbed on a paper disk (8 x 1.5 mm) in a 10 mL sample tube. After evaporation of the solvent, 200 μL of EtOH, 200 μL of 2.5% linoleic acid in EtOH, 400 μL of 5×10^{-2}M phosphate buffer (pH 7.0) and 200 μL of distilled water were added. The mixture in a stoppered sample tube was kept in a sonicator for a few minutes to dissolve the sample out of the paper disk and incubated at 40 °C. At intervals during the incubation, 100 μL of the mixture was taken into a test tube, mixed with 3 mL of 75% EtOH, 100 μL of 30% ammonium thiocyanate and 100 μL of 2×10^{-2}M ferric chloride. After 3 min the absorbance at 500 nm was measured.

2.5. Culture Conditions for Bacterial Strains

All bacterial strains were grown in a medium containing 0.25% poly-peptone, 0.25% yeast extract, 0.01% $FeSO_4 \cdot 7H_2O$, 50% sea water (pH 7.4) at 25°C for 24 h with reciprocally shaking.

2.6. Screening Method for Antioxidant-Producing Microbial Strains (Fig. 9)

Gills, viscera and/or scales of fish and shellfish (*ca.* 2 - 3 g each) were homogenized in sterilized sea water (20 ml). Each suspension was diluted with sterilized sea water from 10^{-1} to 10^{-6}. The suspensions from 10^{-4} to 10^{-6} (0.2 ml) were spread on agar plates made from the medium and cultured at 25°C for a few days. A sterilized filter paper was placed on the agar plate so that colonies and their metabolites were replicated on the paper. Incubation was further continued at 25°C for a few days. Then the filter paper was taken out and sprayed with a DPPH solution (80 mg/ml in EtOH) after drying. Strains showing a white-on-purple spot were regarded as antioxidant-producing bacterial strain. Numbers of the antioxidant-producing strains from each fish or shellfish are listed in Table II.

2.7. TLC Analysis of EtOAc Extracts of Bacterial Fermentation Broths

Active strains were cultured in the liquid medium described above. Each fermentation broth was extracted with EtOAc after being centrifuged. The same amount of each extract was analyzed by TLC: silica gel $60F_{254}$ (Merck) 0.25 mm thickness, benzene-acetone=2:1 and $CHCl_3$-MeOH=10:1. Patterns of spots visualized by spraying the DPPH reagent of each extract were classified into several groups. Typical examples were shown in Fig. 10.

2.8. Comparison of Bacterial Production of Indole between the Aerobic and Less Aerobic Conditions

Two indole-producing strains, No. *Co*-2 from *Conomurex luchuanus* ("Magakigai" in Japanese) and No. *La*-5 from *Lambis lambis* ("Kumogai" in Japanese), were cultured in the liquid medium described above in the Erlen-meyer flasks whose mouths were covered with 4 layers of the elastic plastic film (Parafilm[®]) at 25°C for 24 h with reciprocally shaking. Each EtOAc extract of the fermentation broth was analyzed by HPLC (Nucleosil 50–5; 0.46 x 25 cm; $CHCl_3$-MeOH=20:1; detected at 254 nm) and its indole production was compared with that of the fermentation broth obtained by the normal aerobic cultivation. Results are shown in Fig. 12.

2.9. A Modified Ferric Thiocyanate Method for Evaluation of Antioxidative Activity

The following reagents were prepared: linoleic acid (25 mg/ml in EtOH), NH_4SCN (0.3 g/ml in distilled water), ferrous chloride (2.45 mg/ml in 3.5% hydrochloric acid) and phosphate buffer (5×10^{-2} M $NaH_2PO_4 \cdot 2H_2O$ and 5×10^{-2} M $Na_2HPO_4 \cdot 12H_2O$ were mixed and adjusted to pH 7.0). Each sample solution (200 ml in EtOH) was put into a sample tube (ø 1.5 x 4.3 cm) with a screwed cap together with the linoleic acid solution (200 ml), the phosphate buffer (400 ml) and distilled water (200 ml). In the case of uric acid, EtOH (200 ml) was used for distilled water. The sample tubes were kept at 40°C under the dark conditions. Each sample solution (100 ml) was put into a colorimetric tube together with 75% EtOH (3 ml), the NH_4SCN solution (100 ml) and the ferrous chloride reagent (100 ml described above). About 3 min later, absorbance was measured at 500 nm.

2.10. Isolation and Identification of Antioxidative Compounds (Indole, Uric Acid,3,4-Dimethoxyphenol and 3-Hydroxyindolin-2-one) from Bacterial Isolates

See reference 17.

2.11. Culture Conditions for Fungal Strains

All fungal strains were grown in GP medium containing 2.0 % glucose, 0.5% poly-peptone, 0.2% yeast extract, 0.1% KH_2PO_4 0.05% $MgSO_4\cdot7H_2O$, 50% sea water (pH 6.0) at 25°C with reciprocally shaking.

3. RESULTS AND DISCUSSION

3.1. Screening for Antioxidants from Viscera of Fish and Shellfish

In order to efficiently screen metabolites of marine organisms for antioxidative com-pounds, we designed a simple and time-saving assay procedure[13] as shown in Fig. 1. On a paper disk for an conventional antimicrobial test, 2.0 µg of methyl linolate and each sam-ple extract were absorbed after mixing them well in a microsyringe. After incubating at

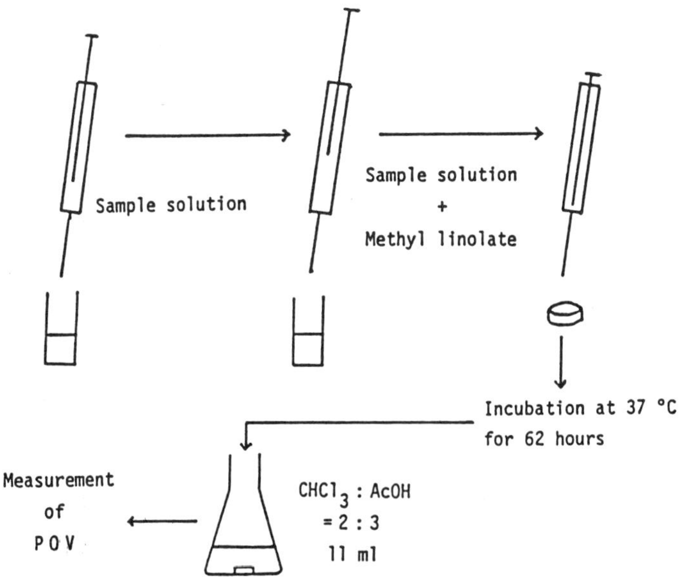

Figure 1. A simple assay procedure for antioxidants using paper disks.

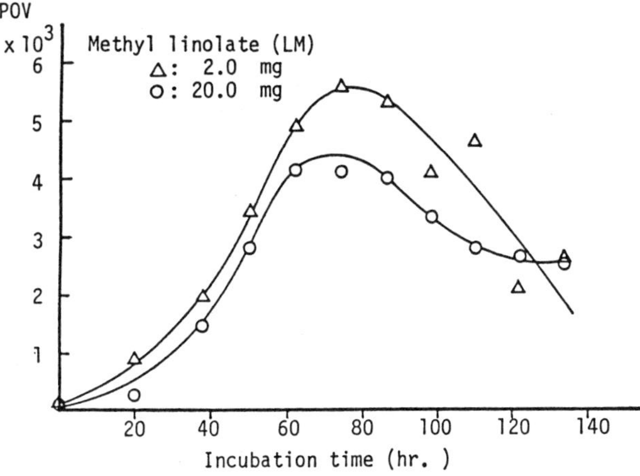

Figure 2. POV and mutagenicity.

37°C, each disk was subjected to the conventional POV measurement and the rec-assay developed by Kada *et al*[15].

Fig. 2 shows the results when 2.0 and 20 mg of methyl linolate alone were applied on the disk. POV and mutagenicity reached maxima after 70–80. So the POV measurement was carried out after 62 hrs in the following screening test. Addition of 1.0 μg of BHT did not give measurable POV.

About 26 kinds of fresh fish and mollusk viscera were collected and extracted following the scheme shown in Fig. 3. Finely chopped viscera was extracted with acetone, EtOH and then 60% EtOH. From the extracts were obtained AH, AEA, EH and EEA fractions. Each fraction was subjected to the aforementioned POV measurement and the rec-assay.

Fig. 4 shows the results of the acetone-hexane fraction, POV of the samples were roughly proportional to the mutagenicity. Some of the samples showed very low activity in the both assay. Especially the bivalves, the short-necked clam, *Ruditapes philippi-*

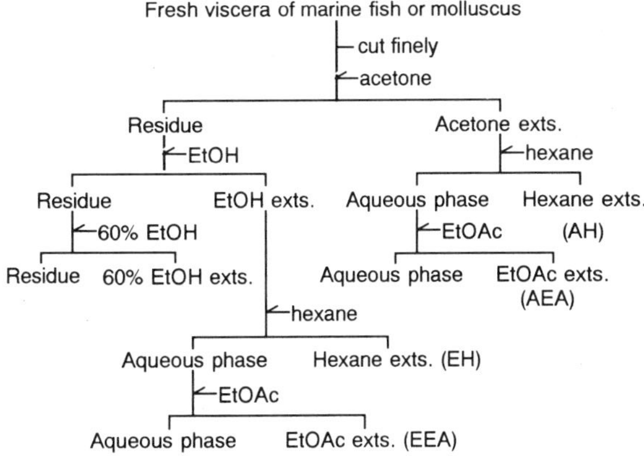

Figure 3. Solvent extractive fractionation of fish and shellfish viscera.

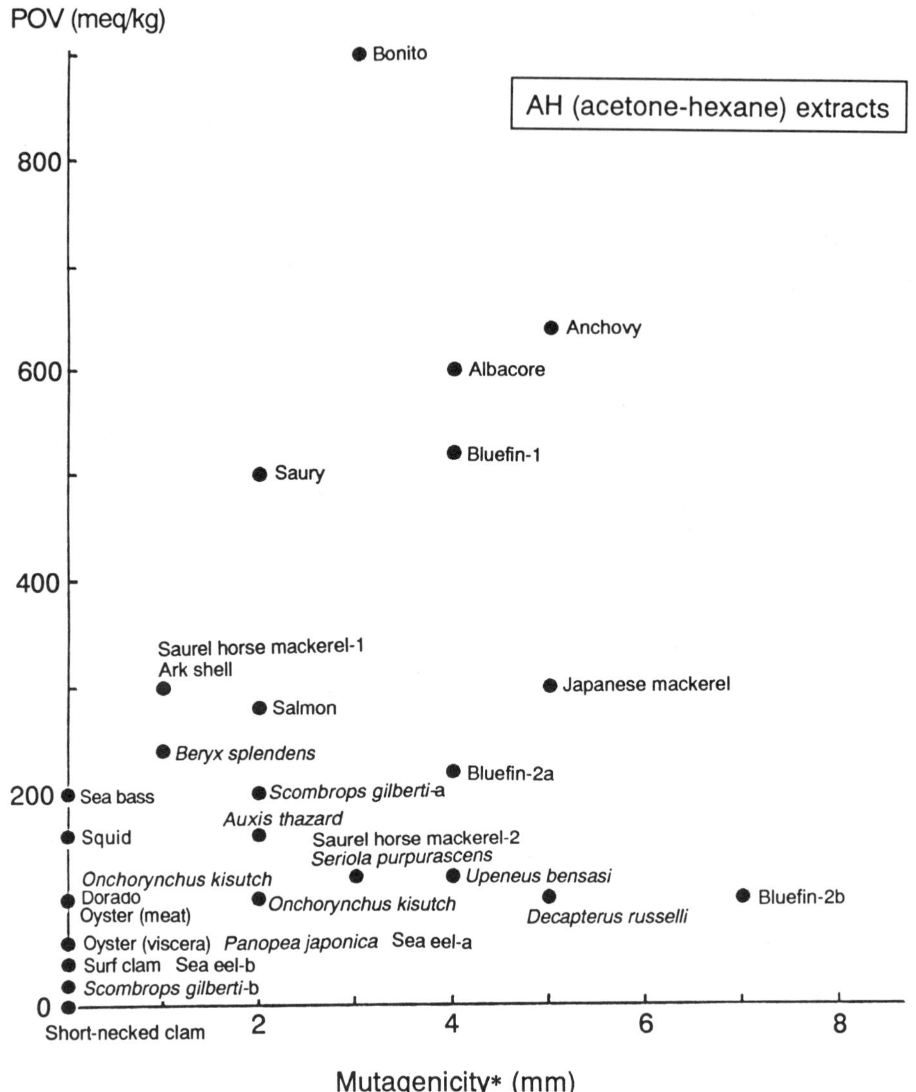

Figure 4. POV and mutagenicity of AH fraction from fish and shellfish viscera.

narum, attracted our attention. Similar results were obtained in the other AEA, EH and EEA fractions.

Some sea-foods like oyster and eel are well known to contain higher amount of tocopherols. So the content of tocopherols in the AH fraction were estimated by HPLC analysis. The extract actually contained very minute amount of α-tocopherol, but the content was not enough to explain the potent antioxidative activity of the crude AH fraction.

Edible part of the clam were extracted with a mixture of $CHCl_3$-MeOH (2:3). Isolation of active principles in the MeOH extract was guided by the antioxidant assay using paper disks to yield 4 active compounds, tentatively named as compounds **1**, **2**, **3**, and **4** (Fig. 5).

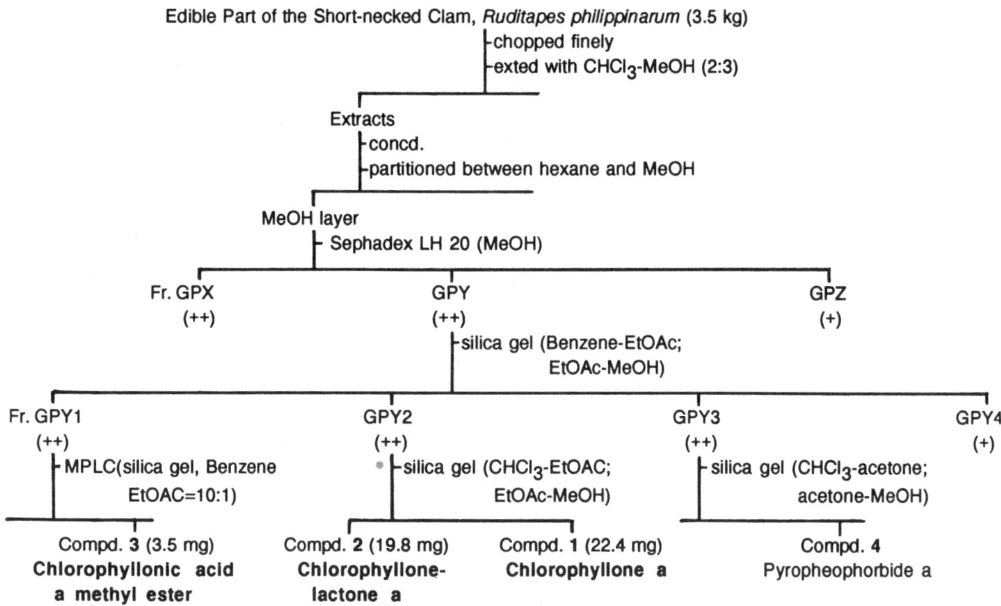

Figure 5. Isolation procedures of chlorophyllone *a* (**1**), chlorophyllonelactone *a* (**2**), chlorophyllonic acid *a* methyl ester (**3**) and pyropheophorbide *a* (**4**).from the short-necked clam, *Ruditapes philippinarum*.

Comprehensive spectroscopic studies especially ^{1}H- and ^{13}C-NMR spectral analyses revealed that new compounds **1**, **2** and **3** isolated from Fr. GPY have unusual chlorin structures (Fig. 6). X-ray crystallographic analysis of **3** confirmed the structure[18]. From the other active fraction, Fr. GPX, compounds **6**, **7** and **8** (Fig. 6) were isolated in the similar manner[13, 14].

Fig. 7 shows the antioxidative activity of chlorophyllone *a* (**1**), chlorophyllonelactone *a* (**2**), chlorophyllonic acid *a* methyl ester (**3**) and pyropheophorbide *a* (**4**) in the ferric thiocyanate method after a slight modification to minimize the sample amount to be tested. They showed potent activity in the modified antioxidant assay. Other chlorin compounds also showed the similar antioxidant activity. Kaneda et al. have already shown that chlorophyll related compounds are photosensitizer under the light[19], but potent antioxidants in the dark[20]. All these compounds also showed higher antioxidative activity than 20 μg of α-tocopherol and about the same level of synthetic antioxidant BHT at doses of 3 μg for 50 μg of linoleic acid.

As compounds **1**, **2** and **3** were new chlorophyll *a* related compounds, we were very interested in the origin of these compounds. So we examined the extracts of viscera of scallops, and oysters which are also plankton feeders. Compound **1** was confirmed to be present in the viscera of the oyster. Compounds **2**, **3**, **4** and **5** were isolated and identified from the scallop[14]. Compound **5** is a new member of this type of compounds, that is, 13^2-epimer of chlorophyllone *a*. It is quite interesting that the scallops contain no chlorophyllone *a*, but its 13^2-epimer. This fact suggests that these compounds are produced by enzymic reaction.

From the facts as shown above we wondered if chlorophyllone *a* (**1**) or its related compounds are produced by microalgae. We obtained a mixture of diatoms mainly composed of *Fragilaria, Nitzschia, Cocconeis and Hyalodiscus* spp. which were grown for

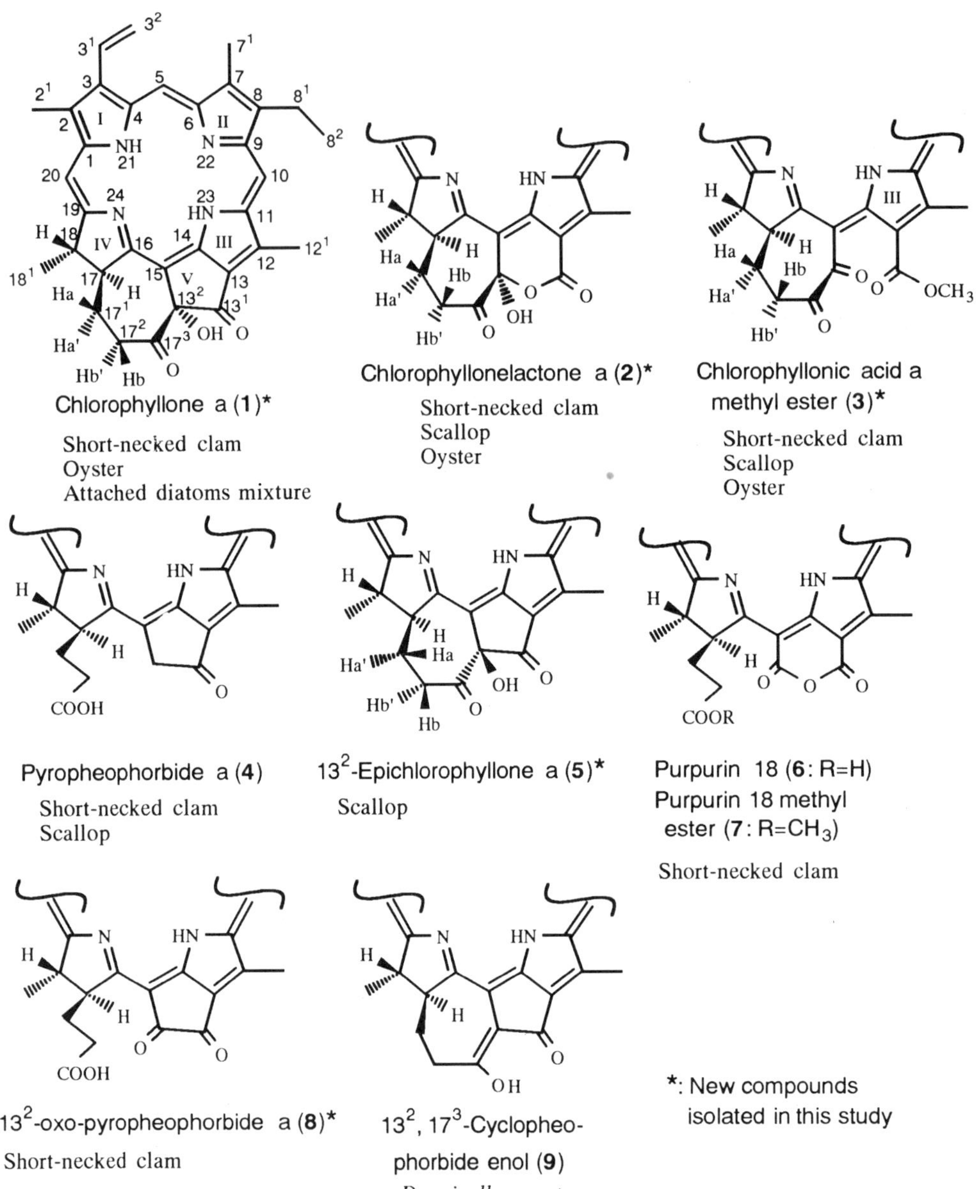

Figure 6. Structures of chlorophyll *a* related compounds and their origin.

seedling production of baby abalone. Chlorophyllone *a* (**1**; 0.6 mg) was isolated and identified from the diatom mixture (dry weight 18 g). This fact indicates that these new type of chlorine compounds were not degradatively produced from chlorophyll *a* in the viscera of the bivalves, but in the diatoms themselves.

Recently Karuso et al. reported the isolation of 13^2, 17^3-cyclopheophorbide enol (**9**) from the sponge, *Darwinella oxeata*[21]. Compounds **1**, **2**, **3**, **5** and **9** belong to a new group

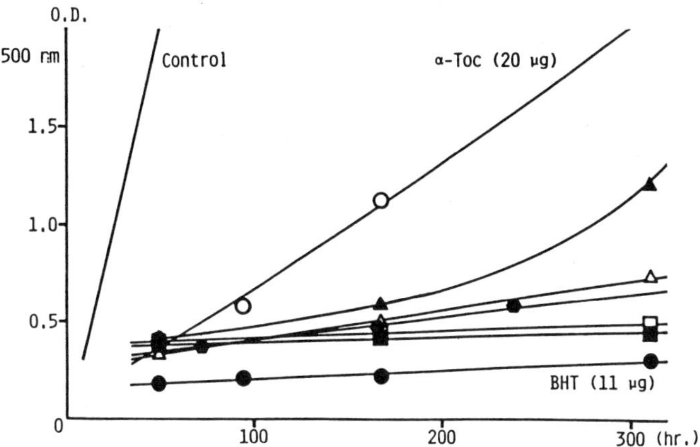

Figure 7. Antioxidative activity of chlorophyllone *a* (**1**), chlorophyllonelactone *a* (**2**), chlorophyllonic acid *a* methyl ester (**3**), pyropheophorbide *a* (**4**) and 13^2-*epi*-chlorophyllone *a* (**5**). Δ, 1(3µg); ■, 2(3µg); □, 3(µg); ●, 4(2µg); Δ, 5 (3µg).

of chlorophyll *a* related compounds and must be produced through a new degradation pathway of chlorophyll *a* as shown in Fig. 8^{13}.

An enolate liberated from pyropheophorbide *a* is cyclized to $13^2,17^3$-cyclopheophorbide enol, which is oxidized and yield chlorophyllone *a* (**1**) and its 13^2-epimer. These compounds can be oxidized to chlorohyllonic acid *a* methyl ester (**3**), which is cyclized to give chlorophyllonelactone *a* (**2**). Another oxidation pathway like Bayer-Viligour oxidation may be involved for the formation of chlorophyllonelactone *a* (**2**).

3.2. Screening for Antioxidants from Microorganisms Originated from Fish and Shellfish Using DPPH

Next our attention was focused into the microorganisms originated from fish and shellfish as sources for antioxidants. We expected that many kinds of microorganisms can be obtained from those marine organisms from various places in the sea. We developed a simple screening method (Fig. 9) for antioxidant-producing strains using 1,1-diphenyl-2-picrylhydrazyl (DPPH) which possess a free radical and shows a characteristic absorption at 517 nm (purple). The purple color rapidly fainted when DPPH encountered with any proton-radical scavengers. Tocopherols, typical natural antioxidants, can be sensitively detected.

Viscera of fish and shellfish and scales and gills of fish were suspended in sterilized sea water and diluted. Aliquot of each suspension was cultured on agar plates containing modified Anderson medium under aerobic conditions. We also expected that bacteria from the viscera of fish and shellfish may produce some antioxidants to adopt themselves to the aerobic conditions. After a few days incubation a sterilized filter paper was placed on the agar plate. After further incubation to replicate the antioxidants produced by each colony, the filter paper was taken out and sprayed with a DPPH solution. A white-on-purple spot showed an antioxidant-producing strain.

Guided by the screening method, we obtained 112 bacterial isolates producing antioxidants from 16 kinds of fish and shellfish: gastropods from the tidal area, bivalves from the shallow water, cuttle fish, eels, a kind of cat fish from the bottom of the sea and yel-

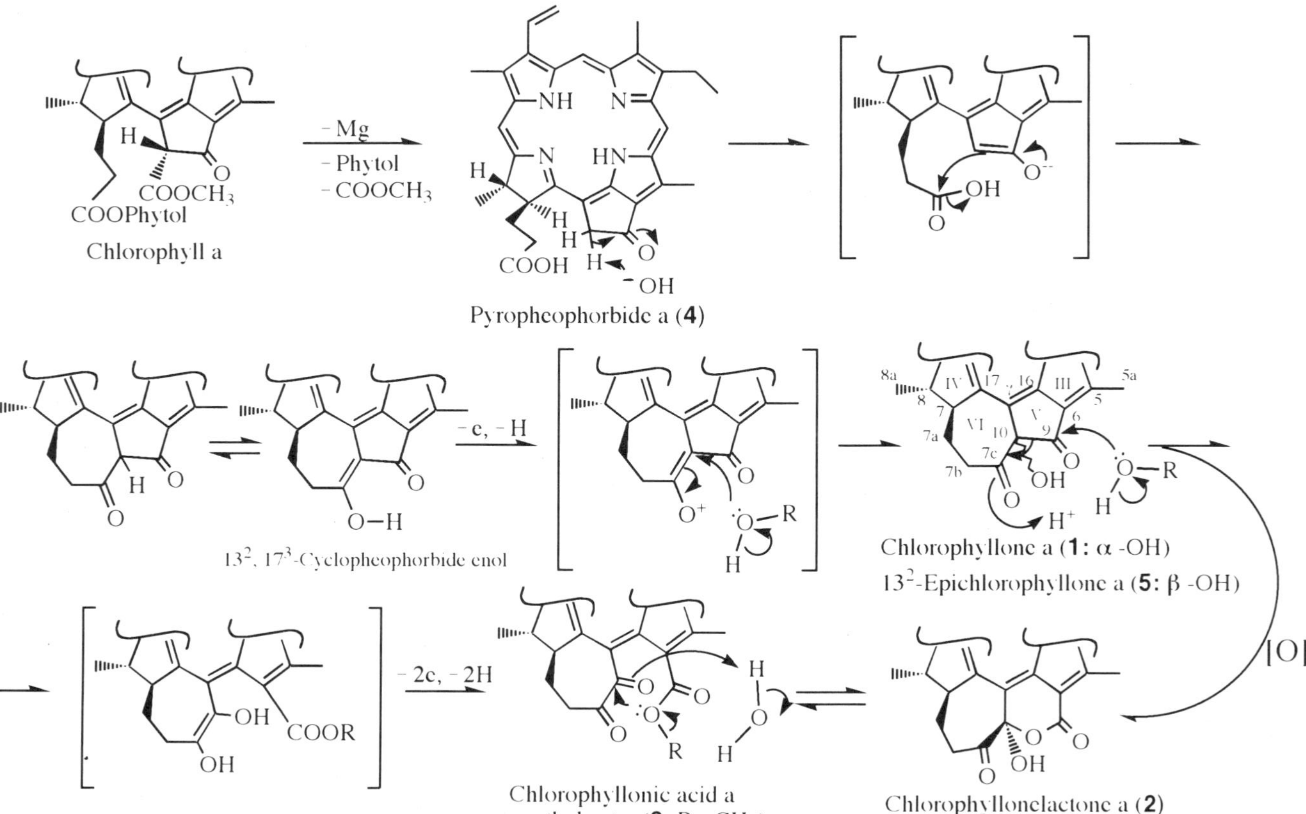

Figure 8. A hypothetical degradation pathway of chlorophyll *a* to chlorophyllone *a* (**1**), chlorophyllonelactone *a* (**2**), chlorophyllonic acid *a* methyl Ester (**3**) and 13²-*epi*-chlorophyllone *a* (**5**).

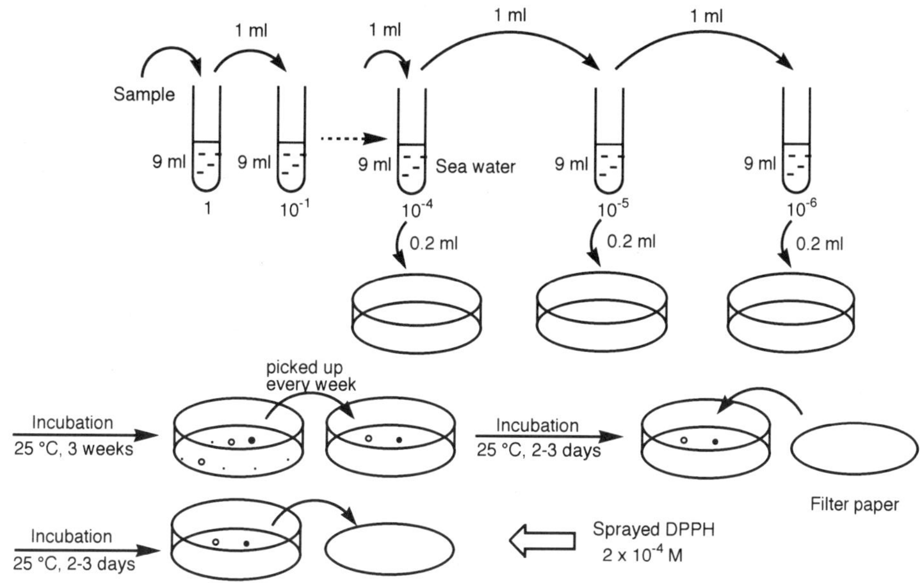

Figure 9. A simple screening method for antioxidant producing microbial strains.

low tails and barracuda from near the surface of the sea (Table II). Bacterial strains were isolated from their appearance and possibly contain the same strains. They were cultured at 25°C for 24 h with shaking under aerobic conditions. EtOAc extract of each fermentation broth was analyzed by TLC.

Typical results were shown in Fig. 10. TLC pattern of each extract is rather simple. Shaded spots were detected by spraying the DPPH reagent. The most typical ones were selected and active principles were isolated and identified. Many of the strains produced indole and 3,4-dimethoxyphenol. The production of indole was remarkable in many of them.

From a bacterial isolate from the short-necked clam *Ruditapes philippinarum* 3,4-dimethoxyphenol and indole were isolated as active principles (Fig. 11). 3,4-Dimethoxyphenol was confirmed to be present in the extracts of bacterial fermentation broths from the eel *Anguilla japonica*, the barracuda *Sphyraena japonica* and the juvenile sardine *Engraulis japonicus* (Fig. 11). 3,4-Dimethoxyphenol is reported as one of the most potent antioxidants among phenolics tested by Zilliken et al.[22] The mechanism of antioxidation of phenolic compounds is considered that a proton radical of a hydroxy group plays a role as a radical scavenger.

From a bacterial fermentation broth from scales of the barracuda *Sphyraena japonica* was isolated 3-hydroxy-indolin-2-one, which was reported to be an intermediate in an oxidative degradation pathway of indole by soil bacteria.[23] There was no description about its chirality. Its CD measurement indicated that the 3-hydroxy-indolin-2-one is a racemate. This is the first report on its antioxidative activity.

The aqueous layer of a fermentation broth of a bacterial isolate from the oyster *Crassostrea gigas* showed potent radical scavenging activity to DPPH. Uric acid was isolated as an active principle. Uric acid is known as an antioxidant in the organisms.[24]

Table II. Origins of antioxidant producing bacteria

Origin	N^a
1. *Trochus maculatus* (Nishikiuzugai)[b]	3
2. *Trochus niloticus* (Sarasabatei)	13
3. *Cypraea vitellus* (Hoshikinutagai)	1
4. *Lunella granulata* (Kangikugai)	1
5. *Lambis lambis* (Kumogai)	6
6. *Turbo marmoratus* (Yakougai)	6
7. *Conomurex luchuanus* (Magakigai)	14
8. *Nerita* (*Amphinerita*) (Ryukyuamagai) *insculpta*	5
9. *Tapes philippinarum* (Asari)	2
10. *Crassostrea gigas* (Magaki)	3
11. *Sepia kobiensis* (Himekouika)	9
12. *Anguilla japonica* (Unagi)	3
13. *Plotosus anguillaris* (Gonzui)[b]	
Gills	4
Viscera	3.
Scales	6
14. *Trachurus japonicus* (Aji)	
Gills	5
Viscera	4
Scales	5
15. *Sphyraena japonica* (Kamasu)	
Gills	5
Viscera	5
Scales	4
16. *Engraulis japonicus* (Shirasu)	5

[a]number of antioxidant-producing strains.
[b]In Japanese

3.3. Indole and Uric Acid Production by Marine Bacteria

Through the isolation of the antioxidants from the bacterial culture, we happened to find that much less amount of indole was produced when the cotton plug was wetted with the culture medium. We confirmed it by the following experiment. Two indole-producing strains (*Co*-2 and *La*-5, respectively) from *Conomurex luchuanus* and *Lambis lambis* were cultured under the aerobic and less aerobic conditions. In the less aerobic case four layers

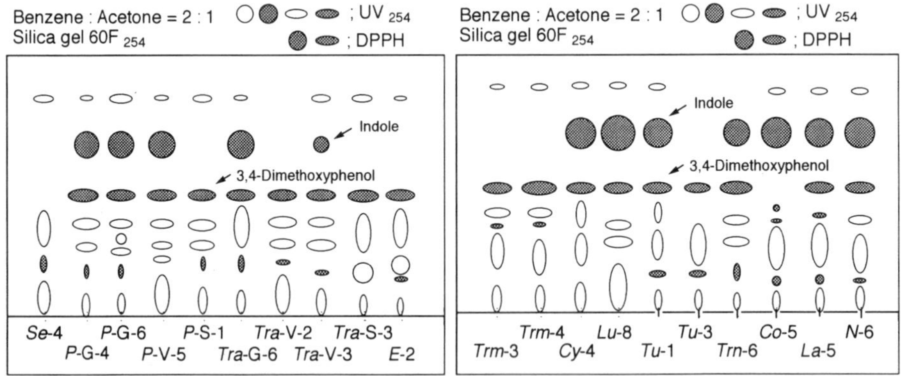

Figure 10. Typical TLC analytical results of EtOAc extracts of fermentation broths of antioxidant-producing bacterial strains from the baby sardine *Engraulis japonicus* (left) and the barracuda *Sphyraena japonica* (right).

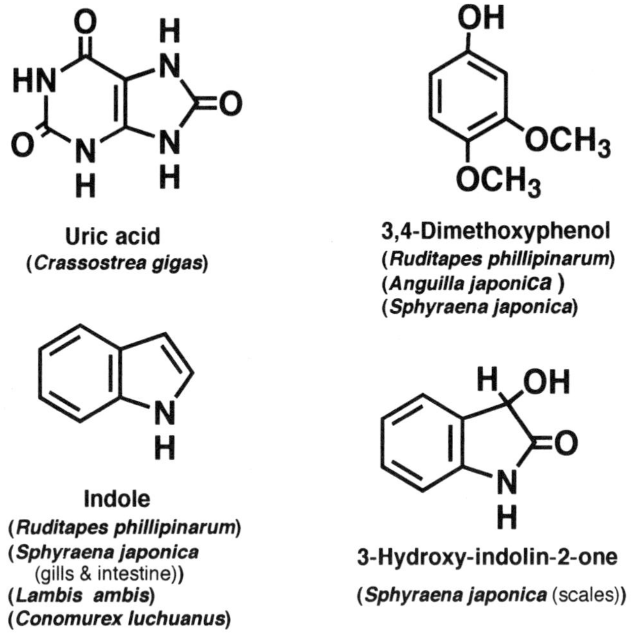

Figure 11. Antioxidative compounds isolated from bacterial isolate originated from various fish and mollusks.

of parafilm were used to seal the mouth of the flask. In both cases, the production of indole dramatically decreased to almost half of it under the less aerobic conditions (Fig. 12) in spite of little difference in the growth between the aerobic and the less aerobic conditions. This may suggest that the abundant production of indole by the bacteria is a kind of adaptation to grow under the aerobic conditions.

The same sort of thing was observed in the case of uric acid. Very little uric acid was produced under the less aerobic conditions. Uric acid is known as an antioxidant in the organisms. So indole may also play an important role as a protector against oxidation in the bacteria which used to live in less aerobic conditions in the viscera of fish or shellfish.

3.4. Antioxidative Activities of the Compounds Isolated from Marine Bacteria

Antioxidative activities of uric acid, indole, 3,4-dimethoxyphenol and 3-hydroxyindolin-2-one were estimated by the ferric thiocyanate method slightly modified by ourselves and compared with those of the conventional antioxidants, BHT and α-tocopherol (Fig. 13). Antioxidative activity of 3-hydroxyindolin-2-on was as strong as uric acid. 3,4-Dimethoxyphenol was inferior to BHT, but showed comparable activity with that of α-tocopherol.

The antioxidative test with DPPH which we used in this experiment is based on a proton radical scavenging action which is one of the various mechanisms of antioxidation. The positive compounds in this test were also positive in the conventional antioxidant test, the ferric thiocyanate method. This simple screening method was found to be efficient to screen antioxidant-producing strains.

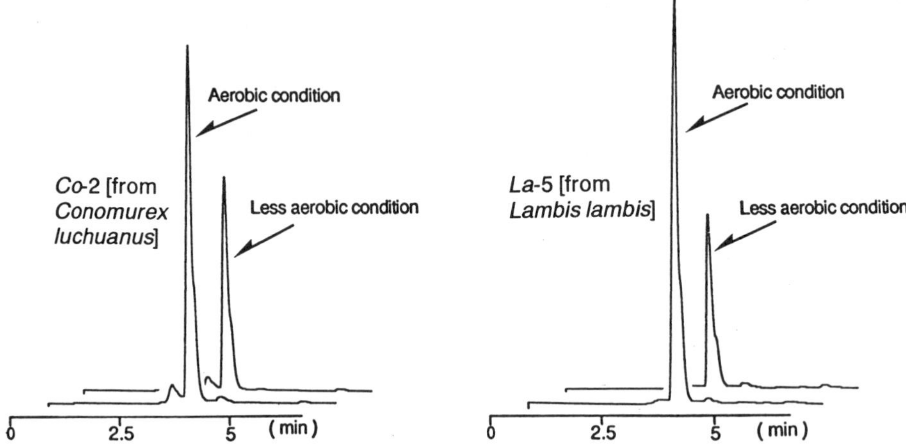

Figure 12.   HPLC analyses of EtOAc extracts of indole-producing strain culture broths. Column: Neucleosil 50-5 (SIL); development: $CHCl_3$-MeOH (20:1), 1.0 ml/min; detection: 254nm.

3.5. Screening for Antioxidants from Marine Fungi

Unfortunately we were not be able to find attractive new antioxidative compounds from marine bacteria. So we next to try to find antioxidants from marine fungi by applying the simple screening method we developed. Fig. 14 shows some typical examples of TLC analysis of the fungal fermentation broth extracts. Fungal metabolites seem to be richer in varieties. We carried out large scale fermentation of some of these strains and isolated a few antioxidative compounds. Results will be shown soon.

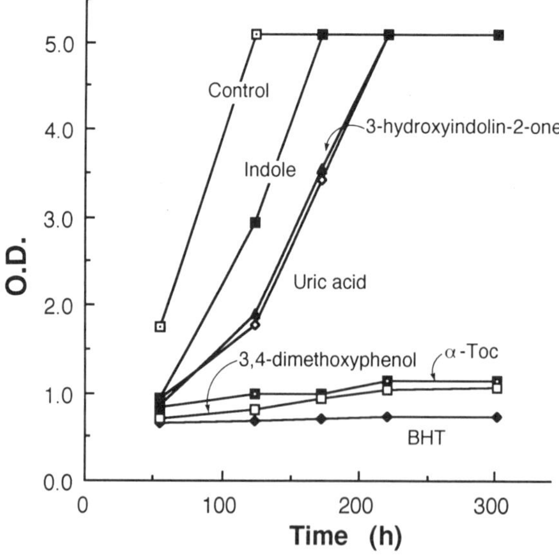

Figure 13.   Antioxidative activities measured by the ferric thiocyanate method.

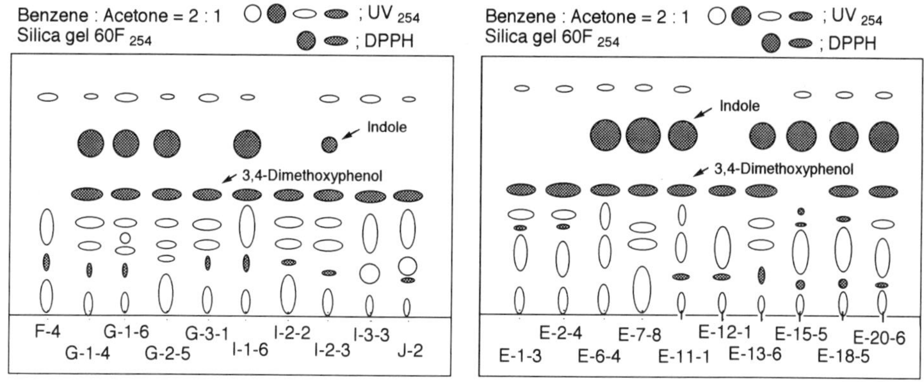

Figure 14. Typical TLC analytical results of EtOAc extracts of fungal fermentation broths from fish and shellfish.

REFERENCES

1. Nakatani, N. (1990)*Nippon Shokuhin Kogyo Gakkaishi*, **37**, 569–576.

2. Larson, R.A. (1988)*Phytochemistry*, **27**, 969–978.

3. Pratt, D.E. and Hudson, B.J.F. (1990) *Food Antioxidants*. ed. B.J.F. Hudson. London: Elsevier. 171–191.

4. Nishibori, S. and Namiki, K. (1985) *Kaseigaku Zasshi*, **36**, 845–850.

5. Fujimoto, K. and Kaneda, T. (1980) *Bull. Japn. Soc. Sci. Fish.*, **46**, 1125–30.

6. Fujimoto, K., Ohmura, H. and Kaneda, T. (1985) *Bull. Japn. Soc. Sci. Fish.*, **51**, 1139–1143.

7. Maruyama, M., Kimura, S., Fujimoto, K. and Kida, N. (1990) Japan Kokai Tokkyo Koho. JP 02245087.

8. Nishibori, S. and Namiki, K. (1988) *Kaseigaku Zasshi*, **39**, 1173–1178.

9. Cahyana, A.H., Shuto, Y. and Kinoshita, Y. (1992) *Biosci. Biotech. Biochem.*, **56**, 1533–1535.

10. Le Tutour, B. (1990) *Phytochemistry*, **29**, 3759–3765.

11. Ishikawa, Y. and Yuki, E. (1975) *Agric. Biol. Chem.*, **39**, 851–855.

12. Yanagimoto, M., Sibazaki, M., Umeda, K. and Kimura, S. (1977) *Nippon Shokuhin Kogaku Kaishi*, **24**, 20–25.

13. Sakata, K., Yamamoto, K. and Watanabe, N. (1994) *Food phytochemicals for cancer prevention II. ACS Symposium Series 547.* Antioxidative compounds from marine oraganisms., eds. C. Ho, T. Osawa, M. Huang, and R.T. Rosen. Washington, DC: American Chemical Society. pp. 164–182.

14. Watanabe, N., Yamamoto, K., Ishikawa, H., Yagi, A., Sakata, K., Brinen, L.S. and Clardy, J. (1993) *J. Nat. Prod.*, **56**, 305–317.

15. Tajima, Y., Kondo, S., Kada, Y. and Sotomura, A. (1980) *Kankyo Hen'igen Jikkenho.* Tokyo: Kodansha.

16. Fukuda, Y., Osawa, T., Namiki, M. and Ozaki, T. (1985) *Agric. Biol. Chem.*, **49**, 301–305.

17. Takao, T., Kitatani, F., Watanabe, N., Yagi, A. and Sakata, K. (1994) *Biosci. Biotech. Biochem.*, **58**, 1780–1783.

18. Yamamoto, K., Sakata, K., Watanabe, N., Yagi, A., Brinen, L.S. and Clardy, J. (1992)*Tetrahedron Lett.*, **33**, 2587–2588.

19. Endo, Y., Usuki, R. and Kaneda, T. (1984)*J. Am. Oil Chem. Soc.*, **61**, 781–784 .

20. Endo, Y., Usuki, R. and Kaneda, T. (1985) *J. Am. Oil Chem. Soc.*, **62**, 1387.

21. Karuso, P., Bergquist, P.R., Buckleton, J.S., Cambie, R.C., Clark, G.R. and Rickard, C.E.F. (1986) *Tetrahedron Lett.*, **27**, 2177–2178.

22. Jha, H.C., Bergmann, C. and Zilliken, F. (1984) *Biochem. Phermacol.*, **33**, 1893–1895.

23. Fujioka, M. and Wada, H. (1968) *Biochem. Biophys. Acta.*, **158**, 70–78.

24. Ames, B.N., Cathcart, R., Schwiers, E. and Hochstein, P. (1981) *Proc. Natl. Acad. Sci. USA*, **78**, 6958–6862.

25. Yokoyama, A., Izumida, H. and Miki, W. (1994) *Biosci. Biotech. Biochem.*, **58**, 1842–1844.

26. Foti, M., Piattelli, M., Amico, V. and Ruberto, G. (1994) *J. Photochem. Photobiol.*, **26**, 159–164.

27. Kobayashi, M., Kakizono, T. and Nagai, S. (1993) *Appl. Environm. Microbiol.*, **59**, 867–873.

9

SYNERGISTIC EFFECT OF SESAME LIGNANS AND TOCOPHEROLS

Kanae Yamashita

Sugiyama-Jogakuen University
17-3 Hoshigaoka-Motomachi, Chikusa-ku, Nagoya 464, Japan

Sesame seed has long been used as a health food to prevent aging [1,2]. But few scientific studies have clarified such effects. Sesame seed is rich not only in oil (about 50%) and protein (about 20%), but also in characteristic lignans (such as sesamin and sesamolin). Empirically sesame oil is known to be highly stable against oxidative deterioration when cooed, and these highly antioxidative properties are apparently related to its characteristic lignan components [3,4].

We conducted three experiments to clarify the anti-aging effects of sesame seed. In first experiment, we examined the effect of sesame seed on senescence by using senescence-accelerated mouse developed by takeda et al.[5]. We have found that the advancement of senescence was apparently suppressed by long-term feeding of sesame seed[6-8].

On the other hand, vitamin E is recognized as possibly having an anti-aging effect. γ-tocopherol is a stronger antioxidative than α-tocopherol in vitro tests, but its vitamin E activity is estimated to be only 5–16% that of α-tocopherol[9]. The tocopherol of sesame seed has been shown to be predominantly γ-tocopherol, with only trace amounts of α-tocopherol. If sesame seed is truly an anti-aging food, there is a question of whether it acts without the contribution of vitamin E activity. Therefore, in the second experiment we examined the relative vitamin E activity of sesame seed, α-tocopherol, and γ-tocopherol in rats. We found that sesame seed showed nearly the same level of vitamin E activity as α-tocopherol because sesame seed lignans and γ-tocopherol act synergistically to produce higher vitamin E activity than γ-tocopherol[10].

This interesting synergism was observed in diets with no α-tocopherol supplementation. However, previous studies have shown that α- and γ-tocopherol are both absorbed similarly from the intestine but that the transporting protein in the liver preferentially bind α-tocopherol, while most γ-tocopherol does not combine with the transporting protein and is excreted through the bile [11-15]. It would be valuable, therefore to clarify whether or not sesame seed lignans act synergistically with α-tocopherol to produce higher vitamin E activity in rats fed a low α-tocopherol-containing diet. Thus, in the third experiment we examined the synergistic effect between α-tocopherol and sesame lignans in rats. We found enhanced vitamin E activity and marked increase of α-tocopherol concentration in the

Food and Free Radicals, edited by Hiramatsu *et al.*
Plenum Press, New York, 1997

blood and tissue of rats fed a low α-tocopherol-containing diet with sesame seed or its lignans. The following account explains these three experiment in more detail.

1. SUPPRESSING EFFECT OF SESAME SEED ON SENESCENCE IN SENESCENCE ACCELERATED MOUSE (SAMP1)

The senescence accelerated mouse (SAM) consists of SAM-p strains that are markedly short-lived and SAM-r strains with normal characteristics of aging. The SAM-p strains grow normally, but then they show early signs aging: severe loss of physical activity and skin glossiness, coarse skin, hair loss, periophtalmic lesions and increased lordokyphosis. Using SAMP-1, one of the SAM p strains we examined the effect of sesame seed or its lignan on senescence in the three experiments.

In experiment 1, the degree of senescence in SAM fed a 20% sesame seed-containing diet was compared with that of a control group fed the standard AIN76 diet[16] containing 10% corn oil. Both diets contained sufficient vitamin E. Sam mice (1.5 months) were given the diets for 7 months and the degree of senescence in the mice was evaluated by a grading score system adopted by hosokawa et al. [17]. **Fig. 1** shows changes in the senescence grading scores of SAM during long-term feeding of black or white sesame seed. With the standard diet, every score increased after 3–4 months feeding (4–5 months in age), while in the case of the sesame seed diet increases in each score were low and suppressed.

In experiment 2, we examined the relation between senescence and the vitamin E activity of sesame seed. Sesame seed contains exclusively γ-tocopherol and negligible α-tocopherol. Because γ-tocopherol has low vitamin E activity[18], sesame seed is a poor source of vitamin E. However, as discussed in the next section, we found that sesame seed lignans and γ-tocopherol act synergistically to produce strong vitamin E activity in rats [10]. It is well known that rats fed a diet deficient in vitamin E exhibit reproductive failure[19]. If sesame seed really has high vitamin E activity, SAM fed a diet containing sesame seed as the only source of vitamin E would show good reproductive success. In the second experi-

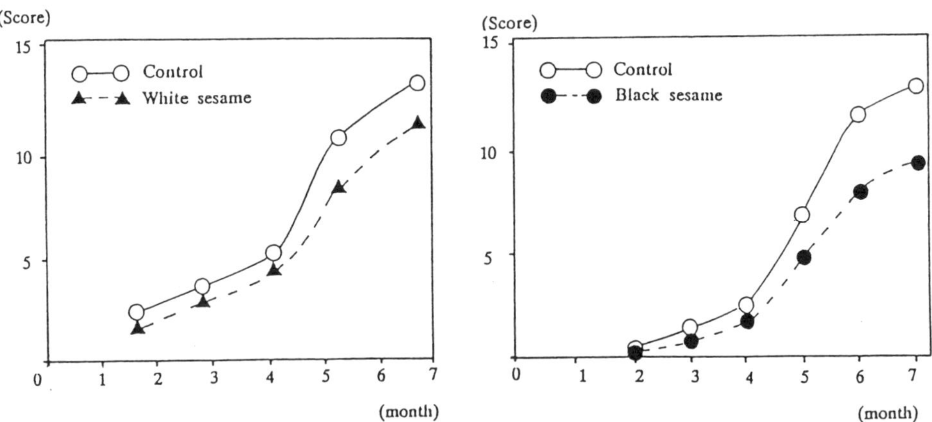

Figure 1. Effect of long term-feeding of sesame seed on changing of senescence grading score in SAMP-1 mice. The mice were given AIN-76 standard diet and 20% white sesame seed (left) or 20% black sesame seed (right) for 7 months.

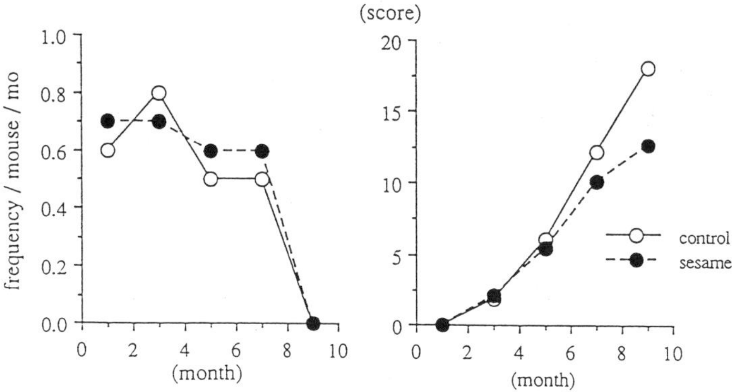

Figure 2. The frequency of delivery (left) and senescence grading score (right) in SAMP-1 mice fed diets containing either sesame seed or α-tocopherol (control) as the vitamin E source.

ment, we studied SAMP1, which were given diets containing either sesame seed or α-tocopherol as the vitamin E source. To investigate the breeding potential, a male-female pair of mice was kept in each cage and the frequency of delivery of mice fed the two diets was recorded over 9 months, while the grading scores of senescence in the male mice were recorded as well. As shown in **fig. 2**, the sesame seed fed-group exhibited nearly the same fertility rate as the α-tocopherol-fed control group. Furthermore, the sesame seed group showed lower senescence grading scores than the α-tocopherol-fed group.

In experiment 3, we studied the suppressive effect of sesaminol on senescence in samp1. Sesame seed contains substantial amounts of lignans such as sesamin and sesamolin (**fig. 3**). There are two types of commercial sesame oil, roasted oil and raw salad oil. The former is extracted from roasted sesame seeds, and the latter extracted from unroasted seeds then refined. Both oils are highly resistant to oxidative deterioration. The

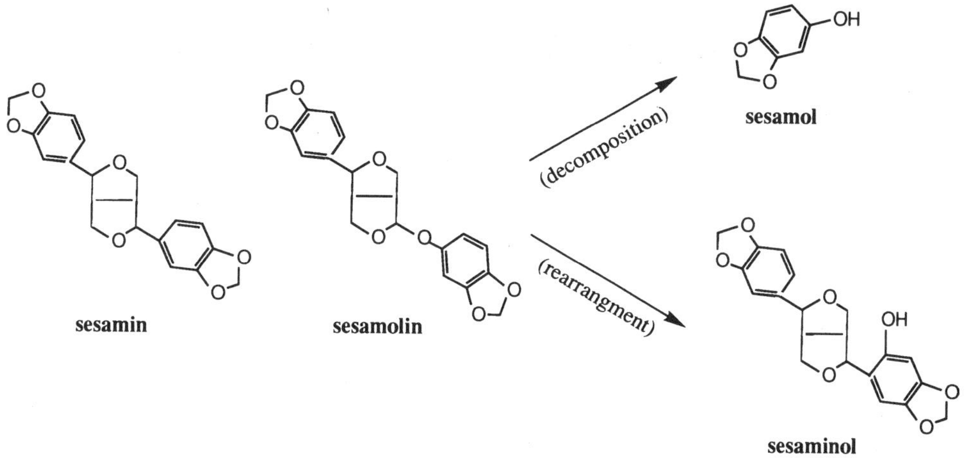

Figure 3. Sesame seed lignans (sesamin and sesamolin) and related antioxidative compounds. (sesaminol). The concentrations of these compounds are as follows: sesamin, 2.5–5.0 mg/g in sesame seed; sesamolin, 1.5–3.0 mg/g in sesame seed; sesaminol 5 mg/g in sesame seed and 0.5–1.0 mg/g in refined raw sesame oil.

antioxidative activity of roasted sesame oil is thought to be caused by the synergistic effects of sesamol formed from sesamolin, γ-tocopherol and unknown maillard reaction products, while that of unroasted oil has been shown to be caused by a new lignan phenol, sesaminol, formed from sesamolin during the refining process of the extracted oil. As reported in section three, in experiments using rats we have observed significantly higher concentrations of α-tocopherol in the plasma and liver of rats fed a low α-tocopherol diet with the addition of sesaminol than in rats fed a low α-tocopherol diet only. In present experiment, we examined whether or not sesaminol and α-tocopherol act synergistically to suppress senescence in sam. Six week old samp1 mice were fed a 50% vitamin E deficient diet (25 mg α-tocopherol/kg diet) with and without 0.1% sesaminol for 20 weeks. **Fig. 4** shows that the senescence scores in SAM were suppressed by the addition of sesaminol, significantly so for hair glossiness and coarseness, and periophthalmic lesions. These three experiments demonstrate the following: 1) sesame seed suppressed senescence in SAMP1. 2) γ-tocopherol in sesame seed exhibited vitamin E activity equal to that of α-tocopherol in suppression of senescence in SAM. 3) sesaminol suppressed senescence in SAM through a synergistic interaction with α-tocopherol.

2. SESAME SEED LIGNANS AND γ-TOCOPHEROL ACT SYNERGISTICALLY TO PRODUCE VITAMIN E ACTIVITY IN RATS

Vitamine is recognized to be a food component that may have an anti-aging properties. γ-tocopherol has stronger antioxidative activity than α-tocopherol in vitro tests, but its vitamin E activity is estimated to be only 5–16% that of α-tocopherol[18]. The tocopherol of sesame seed has been shown to be mostly γ-tocopherol. If sesame seed truly has an anti-aging properties, there is a question of whether it acts without the contribution of vitamin E activity, as well as a question as to the roles of the characteristic lignans of sesame in the anti-aging effects. To this end, we performed two experiments using rats to investigate the effects of sesame seed on vitamin E activity.

In experiment 1, to compare the vitamin E activity of sesame seed relative to that of α-tocopherol and γ-tocopherol, rats were fed for 8 weeks a vitamin-E free diet, an α-tocoppherol-containing diet (51.7 mg/kg diet), a γ-tocopherol-containing diet (51.7 mg/kg diet) or a 20% ground sesame seed-containing diet. The content of γ-tocopherol in the γ-tocopherol -added diet was adjusted to equal that of γ-tocopherol in the ground sesame seed diet. We determined the level of lipid peroxide, oxidative hemolysis and plasma pyruvate kinase activity , which are indices of vitamin E status, in rats fed the experimental diets for 8 weeks. As shown in **fig. 5**, the lipid peroxide concentration of plasma and liver were greater in the control vitamin E-free diet group than in the α-tocopherol-fed group. The γ-tocopherol fed group showed significantly higher peroxide concentration than did the α-tocopherol-fed group, and the sesame seed-fed group had concentrations as low as those of the α-tocopherol-fed group. Hemolysis and pyruvate kinase activity showed results paralleling the peroxide concentrations. Red blood cell hemolysis was as low in the sesame seed-fed group (despite negligible amounts of α-tocopherol in this diet) as it was in the α-tocopherol-fed group. We measured α- and γ-tocopherol concentrations in plasma from rats fed the same experimental diets for 1, 2, 4 and 8 weeks to evaluate changes. The measured concentrations of α-tocopherol showed a relatively high concentration in α-tocopherol-fed rats and, as expected, a relative decline in rats fed the vitamin

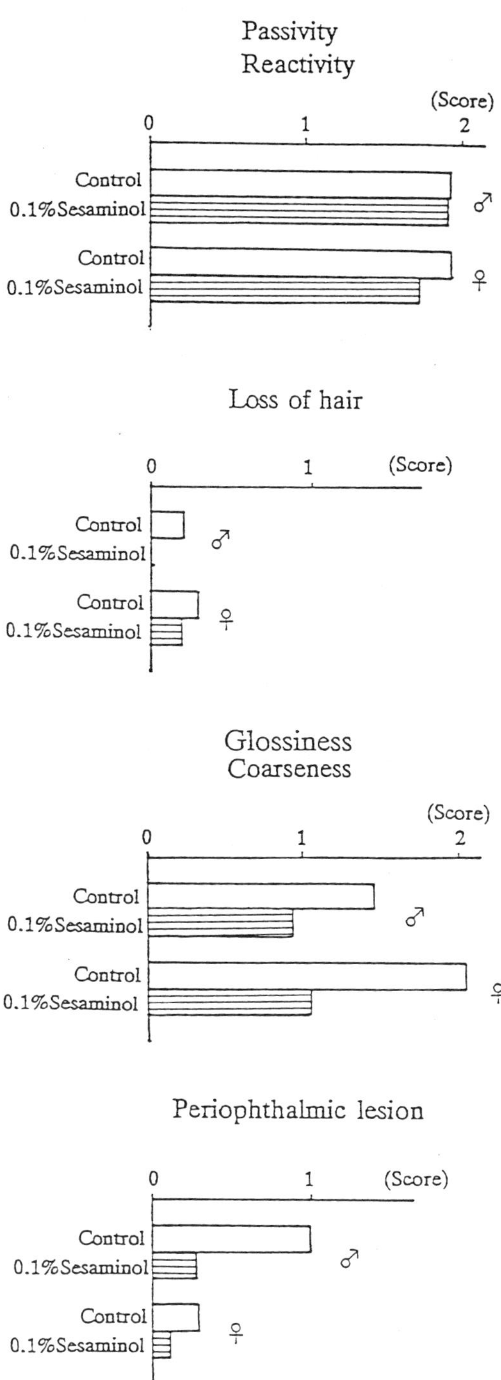

Figure 4. Effect of sesaminol on senescence grading scores of SAMP-1. The mice were fed either diet with vitamin E 50% deficient diets (control) or control plus sesaminol (0.1%) for 20 weeks.

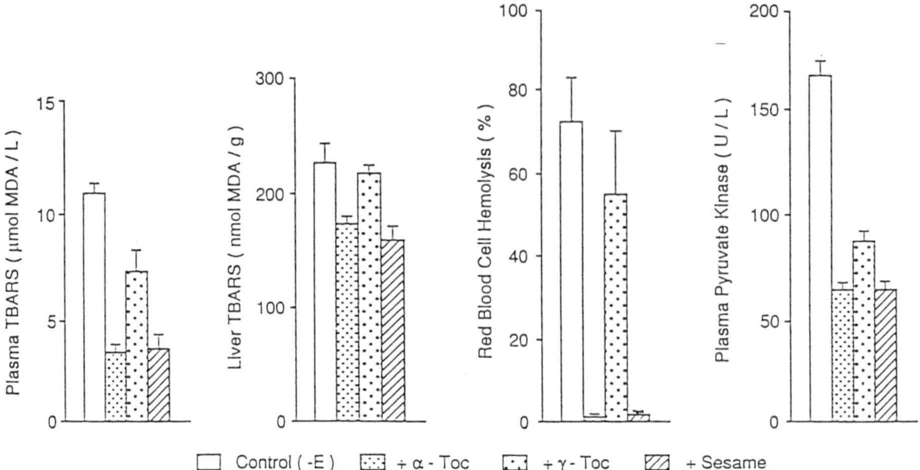

Figure 5. Effect of sesame seed on lipid peroxides in plasma and liver, red blood cell hemolysis and plasma pyruvate kinase activity. Lipid peroxide concentrations were measured by the thiobarbituric acid method, and hemolysis test was performed using dialuric acid.

E-free and γ-tocopherol-containing diets as well as in those fed the sesame seed-containing diet (**fig. 6**). There was a progressive increase in plasma γ-tocopherol concentration in rats fed the sesame seed-containing diet, but no such increase in rats fed the γ-tocopherol-containing diet. The concentration of α- and γ-tocopherol in the liver showed results paralleling the plasma concentration. These results strongly suggest the presence of components in sesame seed that raise the blood and liver concentrations of γ-tocopherol and strengthen the suppressive effect on lipid peroxidation.

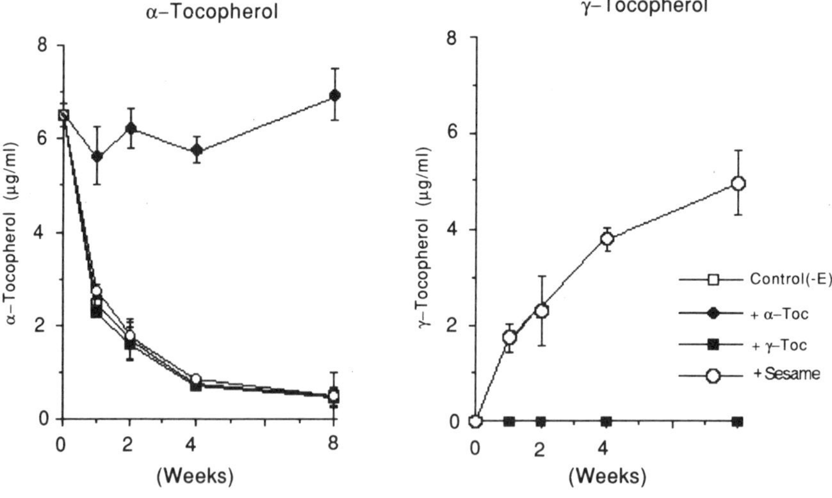

Figure 6. Changing of plasma α- and γ-tocopherol concentrations in rats fed diets contained α-tocopherol, γ-tocopherol or sesame seed as vitamin E source.

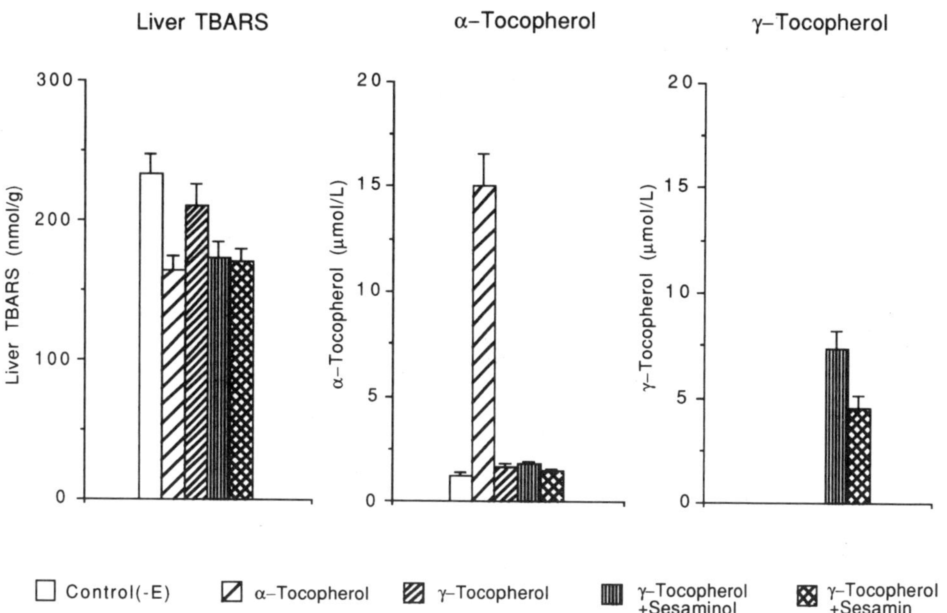

Figure 7. Synergistic effect of sesame lignans with γ-tocopherol on liver lipid peroxide and plasma α- and γ-tocopherol concentrations in rats.

In experiment 2, to examine the synergistic effect of sesame seed lignans with γ-tocopherol, we used sesaminol and sesamin (**fig. 3**). Sesaminol is an antioxidant and sesamin is not. This experiment was performed using a (γ-tocopherol+sesaminol) diet and a (γ-tocopherol+sesamin) diet instead of the sesame seed diet used in the former experiment. For the group fed the vitamin E-free diet, the α-tocopherol-fed group and γ-tocopherol-fed group , the results were essentially the same as in the previous experiment. As shown in **fig. 7,** liver TBARS was significantly low in the α-tocopherol-, the (γ-tocopherol+sesaminol)- and the (γ-tocopherol+sesamin)-fed groups compared to the vitamin E-free and the γ-tocopherol-fed groups. Plasma α-tocopherol concentration was high only in the α-tocopherol-fed group and γ-tocopherol was not found in the γ-tocopherol-fed group, but was present in the (γ-tocopherol+sesamin)-fed group and in significantly greater concentration in the (γ-tocopherol+sesaminol)-fed group. Thus it is evident that both sesaminol and sesamin have the effect of increasing γ-tocopherol concentration in vivo, and the antioxidant sesaminol is superior to sesamin (non-antioxidant). The overall results appear to indicate the sesame seed lignans and γ-tocopherol act synergistically to increase γ-tocopherol concentration and produce high vitamin E activity in rats.

3. MARKED INCREASE OF α-TOCOPHEROL CONCENTRATION IN RATS FED α-TOCOPHEROL CONTAINING DIET WITH SESAME SEED AND ITS LIGNANS

In the preceding experiments the synergistic effects between sesame seed lignans and γ-tocopherol were observed in rats fed diets without α-tocopherol. However, α- and γ-tocopherol are both usually present in an ordinary meal, and α-tocopherol is the most predominant source of vitamin E. Previous studies have shown that the transporting protein in the liver preferentially binds α-tocopherol and most γ-tocopherol is excreted through the bile without combining with the transporting protein[11–15]. Therefore, it would appear to be of value to clarify whether or not sesame seed lignans act synergistically with α-tocopherol to produce higher vitamin E activity. We investigated this subject in the following three experiments.

In experiment 1, the effects of sesame seed on α- and γ-tocopherol concentrations in the plasma and tissue were evaluated in rats fed diets containing no (0), low (10mg/kg diet), normal (50mg/kg diet) or high (250 mg/kg diet) α-tocopherol with or without 20% sesame seed (ca 25 mg γ-tocopherol/100g seed). Rats were divided into 8 groups, each group receiving the experimental diet for 8 weeks. The plasma concentrations of α- and γ-tocopherol are shown in **fig. 8,** indicating that the level of α-tocopherol was higher in proportion to the dietary concentration of α-tocopherol. Further, sesame seed caused a significant increase in α-tocopherol in the plasma of rats fed diets with low, normal, and high concentrations of α-tocopherol, but did not affect the group which was fed no α-tocopherol. On the other hand, γ-tocopherol was detected only in the(no α-tocopherol + sesame seed) and the (low α-tocopherol + sesame seed)-fed groups, but not in either the normal or high α-tocopherol supplemented sesame seed fed groups. This was invariably observed despite the fact that sesame seed contains about 25 mg/100g γ-tocopherol and a negligible amount of α-tocopherol. The concentrations of α- and γ-tocopherol in the red blood cells, liver and kidney paralleled findings in the plasma. Sesame seed caused a significantly higher concentration of α-tocopherol in the plasma and tissue of rats fed diets with three α-tocopherol levels (low, normal or high), as seen in fig 8. According to recent studies using various forms of vitamin E, it is shown that there is no discrimination be-

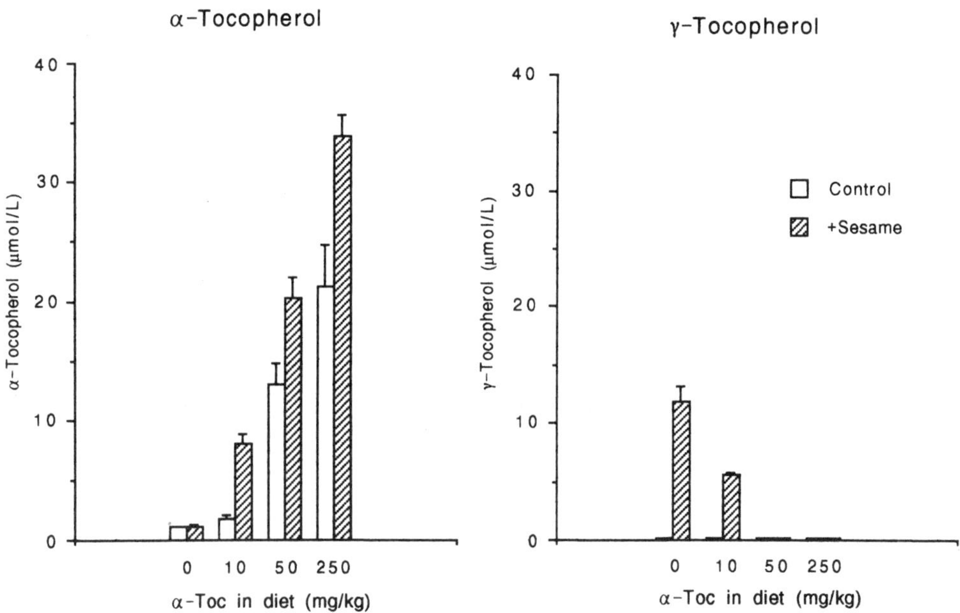

Figure 8. Plasma α- and γ-tocopherol concentrations in rats fed no, low (10 mg/kg), normal (50 mg/kg) and high (250mg/kg) α-tocopherol-containing diets with or without 20% sesame seed.

tween α- and γ-tocopherol during absorption[12, 20, 21], but following uptake by the liver only rrr-α-tocopherol is preferentially bound to tocopherol-binding protein and secreted in nascent vldl[12–14]. Thus, it is suggested that discrimination between forms of tocopherol may occur with the hepatic tocopherol binding protein. In the preceding experiment, we could not detect γ-tocopherol in the plasma and liver of rats fed γ-tocopherol (50mg/kg diet), but γ-tocopherol increased significantly by the addition of sesame lignans to a diet containing the same amount of γ-tocopherol. These results indicate that the binding activity of γ-tocopherol to tocopherol binding protein is weak, as previously reported by Sato *et al.*[13] However, we cannot exclude the possibility that the presence of sesame lignans may enhance the binding activity of γ-tocopherol to the binding protein. These results suggest that the binding activity of tocopherol isomers with tocopherol binding protein is strongly enhanced by the sesame lignans; but it is certain, at any rate, that there is discrimination between α- and γ-tocopherols. The results in experiment 1 may indicate that the supplementation of sesame seed caused an improvement of the vitamin E status of the low α-tocopherol-fed group.

In experiment 2, we examined whether or not the vitamin E status of the low α-tocopherol-fed group was improved by sesame seed supplementation in rats fed low α-tocopherol (10 mg/kg diet) diets supplemented with 5%, 10%, or15% sesame seed, and compared the vitamin E status with the control vitamin E-free diet. We observed that the sesame seed supplemented groups showed extremely low peroxide concentrations in the liver, and even a 5% supplementation with sesame seed to the low α-tocopherol-containing diet almost completely suppressed the high concentration of lipid peroxide in rats fed the low α-tocopherol-containing diet **(fig 9)**. The results of hemolysis and pyruvate kinase activity (not shown) paralleled those of the liver peroxide concentrations. As shown in fig 9, the plasma concentration of α-tocopherol was low in the low α-tocopherol-fed group,

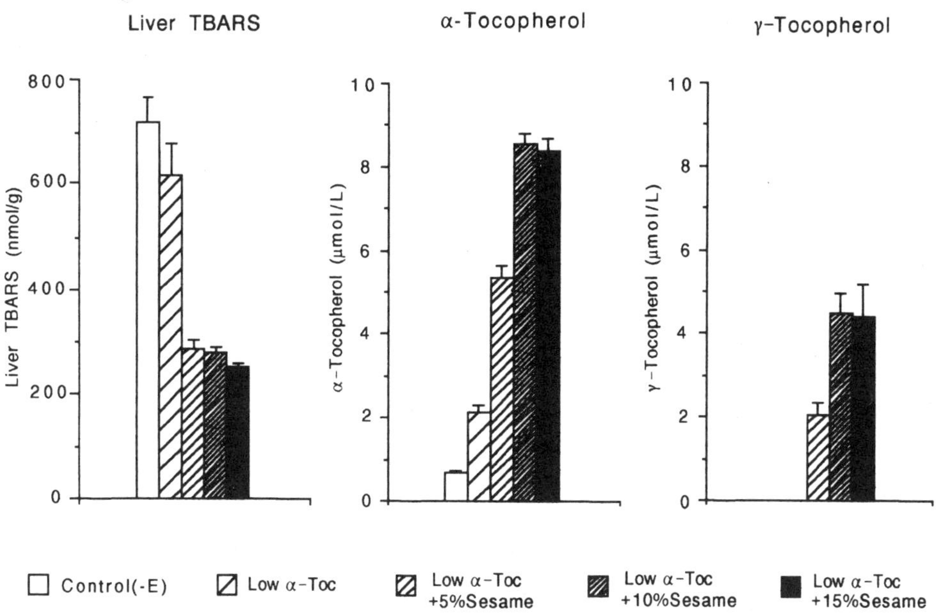

Figure 9. Effect of sesame seed on lipid peroxides in liver and plasma α- and γ-tocopherol concentrations in rats fed low α-tocoherol (10 mg/kg diet) diet supplemented with 5%, 10% and 15% sesame seed for 8 wk.

though it was greater than in the vitamin E-free group; a 5% supplementation of sesame seed to the low α-tocopherol-containing diet caused a significantly higher level of α-tocopherol and γ-tocopherol. The increased tocopherol level was marked with the 10% supplemented group, and no further increase was observed with the 15% supplemented group. The total concentration of α-tocopherol and γ-tocopherol in the plasma in the 10% sesame seed-supplemented group was found to be nearly the same as that of α-tocopherol in the normal α-tocopherol-fed group in the preceding experiment. Although we examined the supplementation effects of sesame seed using a 20% level in that experiment, supplementation with 5–10% sesame seed was shown in experiment 2 to be sufficient to improve the deficient status of vitamin E in the low α-tocopherol-fed group.

Experiment 3 was performed using sesaminol and sesamin instead of the sesame seed used in experiment 2. For the group fed the vitamin E-free diet and the low α-tocopherol-fed group, the results were shown to be essentially the same as in experiment 2. As shown in **fig 10**, low tbars values were observed in the (low α-tocopherol +sesaminol)-, (low α-tocopherol + sesamin)-, and normal α-tocopherol-fed groups, in contrast to the high level found in the low α-tocopherol-fed group. The plasma and liver α-tocopherol concentrations were low in the low α-tocopherol-fed group but slightly higher than in the vitamin E-free group. In the (low α-tocopherol + sesaminol)- and (low α-tocopherol +sesamin)-fed groups, significantly greater concentrations of α-tocopherol were found than in the low α-tocopherol diet with neither sesaminol or sesamin, although the concentration was higher in the sesaminol group than in the sesamin group, and lower in both than in the normal α-tocopherol fed group. Thus it appears evident that both sesaminol and sesamin exhibit the effect of raising α-tocopherol concentrations *in vivo*, and that the antioxidant

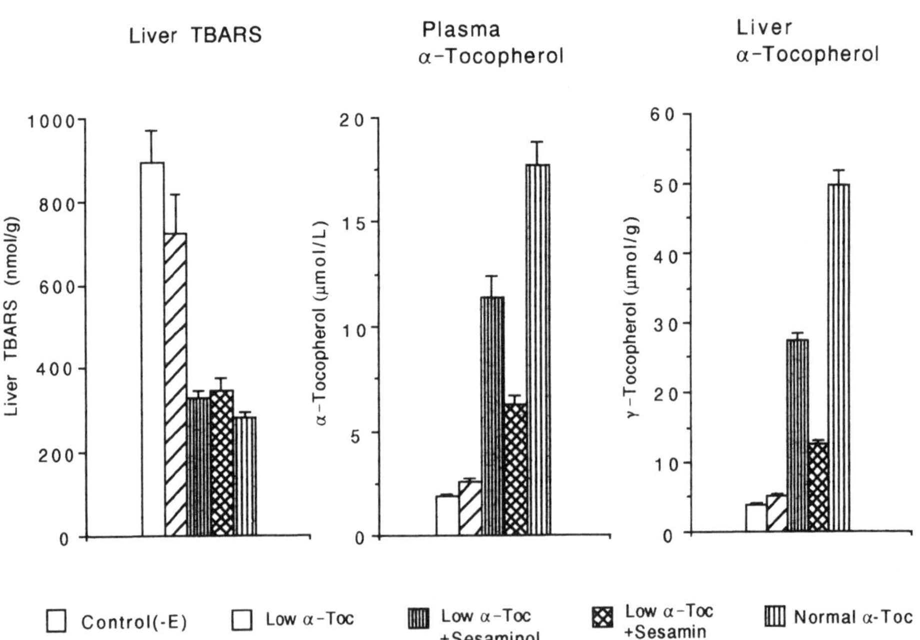

Figure 10. Synergistic effect of sesame lignans with α-tocopherol on lipid peroxides in liver and plasma and liver α-tocopherol concentrations in rats fed low α-tocopherol (10 mg/kg diet) diet supplemented with 2 g sesaminol or sesamin/kg for 8 wk.

sesaminol is superior to sesamin. Sesaminol is mainly found in refined raw salad oil but only in small amounts in sesame seed. Recently katsuzaki et al. [22] reported that there are three new glucosides of sesaminol in defatted sesame cake, and that antioxidative activity is generated by treatment with β-glucosidase. It is thus necessary to further investigate whether or not it is possible for sesamolin, the second major lignan in sesame seed, to be converted to sesaminol *in vivo*.

CONCLUDING REMARKS

In the present experiments, we have found that sesame seed and sesame lignans cause a high α- or γ-tocopherol concentrations in the plasma and tissue of rats and strengthen the suppressive effect of lipid peroxidation. Based on the results of this investigation, we propose that one possible rationale for the consideration of sesame seed as a health food may exist in its enhancing effect on vitamin E activity. However, more thorough experiments are needed to clarify the nature of interactions between sesame lignans and tocopherol-binding protein.

REFERENCES

1. Budowski, P. and Markley, K. S. (1951) The chemical and physiological properties of sesame oil. *Chem. Rev. 48*, 125–151.
2. Namiki, M. (1995) The chemistry and physiological functions of sesame. In *Food Reviews International, 11*, 281–329
3. Fukuda, Y., Nagata, M., Osawa, T. and Namiki, M (1986) Contribution of lignan analogues to antioxidative activity of refined unroasted sesame seed oil. *J. Am. Oil chem. Soc. 63*, 1027–1031.
4. Fukuda, Y., Osawa, T., Namiki, M. and Ozaki, T. (1985) Studies on antioxidative substances in sesame seed. *Agric. Biol. Chem. 49*, 301–306.
5. Takeda, M. G., Hosokawa, M., Takeshita, S., Irino, M., Higuchi, K., Matsushita, T., Tomita, Y., Yasuhira, K., Hamamoto, H., Shimizu, K., Ishii, M. and Yamamuro, T. (1981) A new murine model of accelerated senescence. *Mech. Ageing Dev. 17*, 183–194.
6. Namiki, M., Yamashita, K. and Osawa, T. (1993) Food-related antioxidants and their activities *in vivo*. In *Active Oxygens, Lipid Peroxides and Antioxidants* (Yagi, K., Ed.), pp319–332. Japan Sci. Soc. Press, Tokyo / Crc Press, Boca Raton.
7. Yamashita, K., Kawagoe, Y., Nohara, Y., Namiki, M., Osawa, T. and Kawagishi, S. (1990) Effects of sesame in the senescence-accelerated mouse. *J. Jpn. Soc. Nutr. Food Sci. 43*, 445–449.
8. Yamashita, K. and Namiki M. (1994) Suppressing effect of sesame seed and its lignans on senescence in senescence-accelerated mouse (samp1). In *The SAM Model of Senescence* (Takeda, T., Ed.), pp. 153–156, Excerpta Medica-Amsterdam-London-Newyork-Tokyo
9. Peake, I. R. and Bieri, J. G. (1971) Alpha- and gamma-tocopherol in the rat: *in vitro* and *in vivo* tissue uptake and metabolism. *J. Nutr. 101*, 1615–1622.
10. Yamashita, K., Nohara, Y., Katayama, K. and Namiki M. (1992) Sesame seed lignans and γ-tocopherol act synergistically to produce vitamin E activity in rats. *J. Nutr. 122*, 2440–2446.
11. Catignani, G. L. and Bieri, J. G. (1977) Rat liver α-tocopherol binding protein. *Biochim. Biophys. Acta 497*, 349–357.
12. Kayden, H. J. and Traber M. G. (1993) Absorption, lipoprotein transport, and regulation of plasma concentrations of vitamin E in humans. *J. Lipid Res. 34*, 343–358.
13. Sato, Y., Hagiwara, K., Arai, H. and Inoue, K. (1991) Purification and characterization of the α-tocopherol transfer protein from rat liver. *Febs Lett. 288*, 41–45.
14. Traber, M. G. and Kayden H. J. (1989) Preferential incorporation of α-tocopherol vs γ-tocopherol in human lipoproteins. *Am J. Clin. Nutr. 49*, 517–526.

15. Yoshida, H., Yusin, M., Kuhlenkamp, J., Hirano, A., Stolz, A. and Kaplowitz, N. (1992) Identification, purification, and immunochemical characterization of a tocopherol-binding protein in rat liver cytosol. *J. Lipid Res. 33,* 343–350.

16. American Institute of Nutrition (1977) Report of the American Institute of Nutrition Ad Hoc Committee on Standards for Nutritional Studies. *J. Nutr. 107,* 1340–1348.

17. Hosokawa, M., Kasai, R., Higuchi, K., Takeshita, S., Shimizu, K., Hamamoto, H., Honma, A., Irino, M., Toda, K., Matsumura, M. and Takeda, T. (1984) Grading system: a method for evaluation of the degree of senescence accelerated mouse (sam). *Mech. Ageing Dev. 26,* 91–102.

18. Peake, I. R. and Bieri, J. G. (1971) Alpha- and gamma-tocopherol in the rat: *in vitro* and *in vivo* tissue uptake and metabolism. *J. Nutr. 101,* 1615–1622.

19. Evans, H. M., Bishop, K. S. (1922) The existence of a hitherto unrecognized dietary factor essential for reproduction. *Science 56,* 650–651 (1922)

20. Behrens, W. A. and Madere, R. (1983) Interrelationship and competition of α- and γ-tocopherol at the level of intestinal absorption, plasma transport and liver uptake. *Nutr. Res. 3,* 891–897.

21. Behrens, W. A. and Madere, R. (1987) Mechanisms of absorption, transport and tissue uptake of rrr-α-tocopherol and d-γ-tocopherol in the white rat. *J. Nutr. 117,* 1562–1569.

22. Katsuzaki, H., Kawagishi, S. and Osawa, T. (1994) Sesaminol glucosides in sesame seeds. *Phytochemistry, 35,* 773–776.

10

MIXED NATURAL ANTIOXIDANTS

Midori Hiramatsu

Division of Medical Science
Institute for Life Support Technology
Yamagata Technopolis Foundation
2-2-1 Matsuei, Yamagata 990, Japan

1. INTRODUCTION

Cancer, inflammation, diabetes, cardiovascular disease, neurological diseases and many other diseases have been shown to be related to free radicals. In the case of neurological diseases such as dementia, stroke, and tumor, much effect has been concentrated on the study of free radicals[1]. Recently, food with antioxidant properties has been suggested as a prophylactic for the above diseases. A specially designed mix of natural antioxidants has been recommended as a prophylactic. In the chapter, the free radical scavenging activity of mixed natural antioxidants is described using electron spin resonance (ESR) spectrometry.

2. HERBS

2.1. Sho-Saiko-to-Go-Keishikashakuyaku-to

The Japanese herbal medicine, Sho-saiko-to-go-keishikashakuyaku-to (TJ-960) has been developed as an antiepileptic by Tsumura Co. (Tokyo, Japan). 5 g of TJ-960 includes the extracts from 7 g of Bupreuri Radix, 5 g of Pinelliae Tuber, 3 g of Scutellariae Radix, 4 g of Zizyphi Fructus, 3 g of Ginseng Radix, 2 g of Glycyrrhizae Radix, 1 g of Zingiberis, 6 g of Panelliae Radix and 4 g of Cinnamomi Cortex. TJ-960 has free radical scavenging activity for hydroxyl radicals, superoxide, 1,1-diphenyl-2-picrylhydrazyl (DPPH) as a marker of lipid soluble nitrogen radical, carbon centered radicals and α-tocopheroxyl radical, as determined using ESR spetroscopy[2,3]. In aged rat brain TJ-960 decreased the levels of carbon centered radicals and lipid peroxide[4], increased norepinephrine level in the cerebellum[5], and elevated choline acetyltransferase activity in the hippocampus and striatum[6]. It also improved short-term memory in epileptic patients[7]. These results suggest that TJ-960 may be a prophylactic agent for memory dysfunction with aging.

Food and Free Radicals, edited by Hiramatsu *et al.*
Plenum Press, New York, 1997

TJ-960 inhibited neuronal loss of CA1 area in the hippocampus induced by occluding both common carotid arteries of rats[8] and prolonged the life span of senescence accelerated mice (SAMP8)[9]. In view of these results, TJ-960 may be a suitable medicine for prevention of neurological disorders and aging related to free radicals.

2.2 Baicalein

One of the active components of TJ-960 is probably baicalein, which is focused in Scutellariae Radix. Baicalein scavenged superoxide and DPPH radical as measured using ESR, but did not scavenge hydroxyl radical. It inhibited the loss of pyramidal cells in CA1 areas of the hippocampus of gerbils following bilateral carotid artery occlusion and suppressed the increase of lipid peroxides in iron-induced epileptogenic foci of rats[10]. In comparison with TJ-960, the free radical scavenging action of baicalein is lower.

2.3 Toki-Shakuyakusan

The Japanese herbal medicine, Toki-shakuyaku-san (TJ-23) was developed as a treatment for gynecological disorders by Tsumura Co. (Tokyo, Japan). 7.5 g of TJ-23 contains the extracts from 4 g of Paeoniae Radix, 4 g of Atractylodis Lanceae Rhizoma, 4 g of Alismatis Rhizoma, 4 g of Hoelen, 3 g of Angelicae Radix and 3 g of Cnidii Rhizoma. TJ-23 increases the activity of nicotine acetylcholine receptors in the cerebral cortex of rats[11] and elevates choline acetyltransferase activity in the cerebral cortex and hippocampus of mice[12]. These results suggest that TJ-23 may enhance the function of cholinergic neurons. TJ-23 scavenges hydroxyl radical, superoxide, carbon centered radical and decreases lipid peroxide levels in homogenate of rat cerebral cortex. It decreased lipid peroxide levels and increased mitochondrial superoxide dismutase activity in the cortex, hippocampus and striatum of aged rats. In addition, it lowered the levels of glutamate and glutamine and maintained levels of dopamine and serotonin in the same areas of aged rat brain[13]. As excitoxicity is thought to be related to damage in brain ischemia, epilepsy, and Alzheimer's disease, it is interesting that TJ-23 decreases levels of glutamate and glutamine. Studies indicate efficacy of TJ-23 in Alzheimer's disease[14,15]. In consideration of the results, TJ-23 may be a suitable medicine for Alzheimer's disease.

2.4. Ganoderma Lucidum (Fr.) Karst

Ganoderma Lucidum (Fr.) Karst has been used to retard aging and for health. Dried Ganoderma Lucidum from Shin Nikko-Kosan Company (Tokyo, Japan) was extracted with boiling water and its free radical scavenging activity was examined. It scavenged superoxide and DPPH radicals and had a weak scavenging effect for hydroxyl radical[16].

GEO REISHI 103, produced by ARSOA Ohsho Corporation (Tokyo, Japan), is a health food containing the extract of Ganoderma Lucidum. It scavenged superoxide and DPPH radicals but increased the generation of hydroxyl radicals[17].

2.5. Kanglaojianshenye

Kanglaojianshenye (KLJSY) is a liquid used to retard aging and for several health in China. 10 ml contains Glymadenia conopsea, Cistanshe salsa and Psoralea corylifolia. It scavenged hydroxyl radical, superoxide, DPPH radical and carbon centered radical induced by auto-oxidation in rat cerebral cortex homogenate[18].

2.6. Long-Life SOD

Long-life SOD is a liquid used to treat senility, produced by Guiyang Long Fa Health-protective Medicine Factory (China). It contains 10 ml of squeezed cili (Ribes burejense Fr. Schmidt). Cili contains vitamins C and E and superoxide dismutase, and its efficacy for antidecrepitude, strengthening immune function of cells, and treating hypertension, diabetes mellitus, jaded appetite, ischemia and constipation has been shown. Long-life SOD has free radical scavenging activity against hydroxyl radicals, superoxide and DPPH radicals[19].

2.7. Fructus Momordicae

Fructus Momordicae from China has anti-inflammatory action. The water extract of Frucuts Momordicae has free radical scavenging action against hydroxyl, superoxide and DPPH radicals. It also inhibits lipid peroxide formation in rat brain homogenate. These results suggest that the anti-inflammatory action of Fructus Momordicae may be due to its antioxidant activity[20].

2.8. Summary of Antioxidant Capacities

No scavenging action against hydroxyl radicals was found in ganoderma lucidum (Fr.) Karst. Other herbs scavenged the water- soluble hydroxyl radical and superoxide radical and the lipid- soluble DPPH radical.

Vitamin E is a lipid-soluble antioxidant and vitamin C is a water-soluble antioxidant. Electrons transfer to vitamin E from vitamin C and from other water-soluble antioxidants, for example, from glutathione, and the Japanese herbal medicine, TJ-960[3]. TJ-960 and vitamin C inhibited the decrease of vitamin E concentration induced by oxidative stress produced with arachidonic acid and lipoxygenase, in vitamin E-enriched rat liver microsomes and submitochondrial particles[3].

Recently, evidence for vitamin E recycling by thiols, such as, dihydroxylipoic acid and glutathione, and by vitamin C cycle, has been reported[21]. For this reason a mix of antioxidants has been suggested as optimum for health.

3. DESIGNED FOOD

Green tea contains a complex mixture of substances, including (-)-epicatechin, (-)-eicatechin gallate, (-)-epigallocatechin and (-)-epigallocatechin gallate. These catechins have free radical scavenging activity[22] and antioxidant actions[23] and their efficacy was shown in cancer[24], hypertension[25] and allergy[26]. "βCATECHIN" was designed as an antioxidant drink and contains a mix of natural antioxidants: green tea extract, ascorbic acid, sunflower seed extract, fat-soluble dunaliella carotene and natural vitamin E. ESR study showed that "βCATECHIN" has free radical scavenging activity against hydroxyl radical, superoxide radical and lipid soluble, 1,1-diphenyl-2-picrylhydrazyl radicals, and inhibits lipid peroxidation in rat brain homogenate and iron-induced lipid peroxidation in the rat brain cortex[27]. Some of its components appear to cross the blood brain barrier and "βCATECHIN" may have benefit as a prophylactic for neurological diseases related to aging which involve free radicals.

4. SUMMARY

Herbal preparations are often mixes of extracted plants, and contain many water- and lipid-soluble components. These antioxidants may help maintain and enhance the body's natural antioxidants. Recently, Alzheimer dementia has been reported to be related with free radicals. "Designed foods" containing natural water- and lipid-soluble antioxidants, which easily cross the blood brain barrier, may prove as valuable prophylactics for neurological diseases, including dementia, related to the action of free radicals.

ACKNOWLEDGMENTS

We thank Dr. Eric Witt for assistance in preparation of this chapter.

REFERENCES

1. L. Packer, L. Prilipko and Y. Christen (eds), Free Radicals in the Brain — Aging, Neurological and Mental Disorders (1992), Springer-Verlag, Berlin/Heidelberg, Germany.

2. Hiramatsu, M., Edamatsu, R., Kohno, M. and Mori, A. (1988) Scavenging of free radicals by Sho-saiko-to-go-keishi-ka-shakuyaku-to (TJ-960). In: E. Hosoya and Y. Yamamura (eds), Recent Advances in the Pharmacology of KAMPO MEDICINES, pp. 120–127. Excerpta Medica, Tokyo.

3. Hiramatsu, M., Velasco, R.D. and Packer, L. (1990) Vitamin E radical reaction with antioxidants in rat liver membranes. Free Rad. Biol. Med. 9, 459–464.

4. Hiramatsu, M., Edamatsu, R., Kabuto, H. and Mori, A. (1988) Effect of Sho-saiko-to-go-keishi-ka-shakuyaku-to (TJ-960) on monoamines, amino acids, lipid peroxides, and superoxide dismutase in brains of aged rats. In: E. Hosoya and Y. Yamamura (eds), Recent Advances in the Pharmacology of KAMPO MEDICINES, pp. 128–135. Excerpta Medica, Tokyo.

5. Hiramatsu, M., Kabuto, H. and Mori, A. (1986) Effect of Shosaikoto-go- keishikashakuyakuto (TJ-960) on brain catecholamine level of aged rats. IRCS Med. Sci. 14, 189–190.

6. Hiramatsu, M., Haba, K., Edamatsu, R., Hamada, H. and Mori, A. (1989) Increased choline acetyltransferase activity by Chinese herbal medicine Shosaiko-to-go-keishi-ka- shakuyaku-to in aged rat brain. Neurochem. Res. 14, 249–251.

7. Nagakubo, S., Niwa, S., Kumagai, N., Fukuda, M., Anzai, N., Yamauchi, T., Aikawa, H., Toyoshima, R., Kojima, T., Matsuura, M., Ookubo, Y. and Ootaka, T. (1993) Effects of TJ- 960 on sternberg's paradigm results in epileptic patients. Jpn. J. Psychiat. Neurol. 47, 609- 620.

8. Sugimoto, T., Ishige, A., Sudo, K., Sekiguchi, K., Iizuka, S., Itoh, K., Yuzurihara, M., Aburada, M., Hosoya, E. and Sugaya, E. (1988) Protective effect of Sho-saiko-to-go- keishika-shakuyaku-to (TJ-960) against cerebral ischemia. In: E. Hosoya and Y. Yamamura (eds), Recent Advances in the Pharmacology of KAMPO MEDICINES, pp. 112–112. Excerpta Medica, Tokyo.

9. Hiramatsu, M., Komatsu, M. and Ueda, Y. Aging and herbal antioxidants. In: W. Xin, L. Packer and F. Gu (eds), Natural Antioxidants: Molecular Mechanisms and Health Effects, AOCS Press, in press.

10. Hamada, H., Hiramatsu, M. and Mori, A. (1993) Free radical scavenging action of bicalein. Arch. Biochem. Biophys. 306, 261–266.

11. Hagino, N. and Koyama, T. (1988) Stimulation of nicotine acetylcholine receptor synthesis in the brain by Toki-shakuyaku-san (TJ-23). In: E. Hosoya and Y. Yamamura (eds), Recent Advances in the Pharmacology of KAMPO MEDICINES, pp. 144–149. Excerpta Medica, Tokyo.

12. Toriizuka, K. (1995) Effect of Toki-shakuyaku-san on central nervous system, memory and immune reaction. Modern Physician 15, 376–383.

13. Ueda, Y., Komatsu, M. and Hiramatsu, M. Free radical scavenging activity of Japanese herbal medicine, Toki-shakuyaku-san (TJ-23), and TJ-23 increased superoxide dismutase activity and decreased concentration of lipid peroxide, glutamate and monoamine metabolites in aged rat brain. Neurochem. Res. in press.

14. Kudo, T. and Sugiura, K. (1992) Toki-shakuyaku-san and aged dementia. Iyaku Journal 28, 35–38 (in Japanese).

15. Totsuka, S. and Kawakatsu, S. (1991) The treatment of Toki-shakuyaku-san on aged dementia. Gendai-Toyo-Igaku 12, 315–317 (in Japanese).

16. Liu, J., Hiramatsu, M. and Mori, A. (1991) Free radical scavenging activity in Ganoderma Lucidum (Fr.) Karst. The Clinical Report 25, 317–321 (Abstract in English).

17. Hohjoh, T., Hiramatsu, M., Komatsu, M. and Takagi, N. (1995) Free radical scavenging action of GEO REISHI 103. Magnetic Resonance in Medicine 6, 312–314.

18. Liu, J., Wang, Z.X., Hiramatsu, M. and Mori, A. (1992) Free radical scavenging activity of Kanglaojian-shenye— A possible mechanism of its anti aging effect. Magnetic Resonance in Medicine 3, 134–137 (Abstract in English).

19. Komatsu, M. and Hiramatsu, M. (1995) Free radical scavenging activity of Long-life SOD. Clinical report 29, 69–74 (Abstract in English).

20. Liu, J. and Mori, A. (1993) Free radical scavenging and antiperoxidative effect of Frucutus Momordicae. The Clinical Report 27, 257–263.

21. Packer, L. (1995) Nutrition and biochemistry of the lipophilic antioxidants vitamin E and carotenoids. In: A.S.H. Ong, E. Niki and L. Packer (eds), Nutrition, Lipids, Health, and Disease, pp. 8–35. AOCS Press, USA.

22. Uchida, S., Edamatsu, R., Hiramatsu, M., Mori, A., Nonaka, G., Nishioka, I., Niwa, M. and Ozaki, M. (1987) Condensed tannins scavenger active oxygen free radicals. Med. Sci. Res. 15, 831–832.

23. Matsuzaki, T. and Hara, Y. (1985) Antioxidative activity of tea leaf catechins. J. Agr. Chem. Soc. Jap., 59, 129–134.

24. Hara, Y., Matsuzaki, S. and Nakamura, K. (1989) Anti-tumor activity of tea catechins. J. Jpn. Soc. Nutr. Food Sci., 42, 39–45.

25. Hara, Y. and Tonooka, F. (1990) Hypotensive effect of tea catechins on blood pressure of rats. J. Jpn. Soc. Nutr. Food Sci., 43, 345–348.

26. Maeda, Y., Yamamoto, M., Matui, T., Sugiyama, K., Yokota, M., Nakagomi, K., Tanaka, H., Takahashi, T. and Kobayashi, T. (1989) Inhibitory effect of tea extract on histamine release from mast cells. J. Food Hyg. Soc. 30, 295–299.

27. Yoneda, T., Hiramatsu, M., Sakamoto, M., Togasaki, K., Komatsu, M. and Yamaguchi, K. (1995) Antioxidant effects of "β-CATECHIN". Biochem. Mol. Biol. Int. 35, 995–1008.

11

RIBOFLAVIN-SENSITIZED SINGLET OXYGEN FORMATION IN MILK

Lawrence J. Berliner and Tateaki Ogata

Department of Chemistry
The Ohio State University
120 W. 18th Ave, Columbus, Ohio 43210

1. INTRODUCTION

The photosensitized oxidation of foods is initiated by natural pigments or sensitizers producing an excited state which may follow Type I or Type II reaction pathways[1]. The interaction of sensitizer with food components such as protein or fat, generating free radicals, is called Type I; The reaction of excited sensitizer producing singlet oxygen is called Type II [2]. Dairy products are excellent sources of riboflavin[3]. Photochemical reactions of riboflavin and various flavin compounds have been studied in several systems[4,5,6]. Riboflavin has been shown to favor a Type I mechanism because it is easily oxidized or reduced and its high water solubility minimizes its interaction with oxygen which is readily more soluble in lipids[7]. The triplet state of riboflavin has been suggested to generate singlet oxygen in milk during exposure to light[8]. ESR can detect the formation of singlet oxygen by reaction with 2,2,6,6-tetramethyl-4-piperidone (TMPD), forming a stable nitroxide radical adduct, 2,2,6,6-tetramethyl-4-piperidone-1-oxyl (TAN or TEMPONE)[9,10].

2. MATERIALS AND METHODS

2.1 Materials

Riboflavin, methanol, 2,2,6,6-tetramethyl-4-piperidone mono-hydrate (TMPD), and 2,2,6,6-tetramethyl-4-piperidone-1-oxyl (TAN or TEMPONE) were obtained from Aldrich Chem. Co. Superoxide dismutase (SOD) from bovine erythrocytes was purchased from Sigma Chem. Co. Whole and skim milk were obtained from a local grocery. Other chemicals used were of reagent grade.

2.2 Photosensitizing Solutions

The spin trapping agent, TMPD, was added to either 5, 10 or 15 µM riboflavin to a final concentration of 20 mM in 25 mM phosphate buffer, pH 7.0. TMPD (20 mM) was

Food and Free Radicals, edited by Hiramatsu *et al.*
Plenum Press, New York, 1997

also added to whole and skim milk samples. The production of TAN in buffer solution and skim milk samples were linearly dependent on TMPD concentration. 20 mM TMPD was chosen for maximum sensitivity of detection. The samples were bubbled with oxygen in a 200 ml beaker at a flow rate of 100 ml/min.under a 15 W fluorescent lamp.

2.3 ESR Instrumental Conditions

Signals were recorded at room temperature on a Varian E-109 X-band ESR spectrometer. Conditions were: microwave power, 20 mW; magnetic field, 3390 ± 50 G; 100 kHz modulation amplitude, 2 or 4 G.

3. RESULTS AND DISCUSSION

An ESR spectrum of TAN formation after 10 minutes of illumination is shown in Figure 1. The time course of formation of TAN in phosphate buffer or milk containing 5 µM riboflavin and 20 mM TMPD or milk (containing containing 20mM TMPD) is shown in Figure 2. In the buffer sample, TAN concentration increased over 10 minutes of illumination and then decreased. When solutions of either: riboflavin (5 µM) and TAN (0.2 µM); TMPD (20 mM) and TAN (0.2 µM); or 20 mM TMPD alone were oxygenated and exposed to light, no changes in the TAN signal intensity occurred. However, TAN increased slightly after 5 µM riboflavin was reoxygenated in the dark, strongly suggesting the ability of bleached riboflavin to reduce the nitroxyl radical to the hydroxylamine[11]. This decay of TAN suggests theTAN generation in oxygenated skim and whole milk samples containing 20 mM TMPD occurred after 5 minutes of illumination. The TAN concentration in skim milk was maximum at 15 minutes of illumination and also then gradually decreased.

The mechanism of signal decay of TAN in milk appears to be different than that observed in riboflavin alone as suggested by the decrease of TAN after addition of 10 µM TAN to milk during light exposure and dark storage (data not shown). The TAN concentration rapidly decreased after 15 minutes, although a TAN sample stored in the dark showed almost no decrease. The signal also did not change when 5 µM riboflavin was added to TAN in 25 mM phosphate buffer. Yet TAN decreased in milk containing riboflavin, which may be indicative of a termination reaction of the nitroxyl compound with other milk components. Further evidence of riboflavin involvement of singlet oxygen formation in milk came from results, obtained by first removing riboflavin from milk by

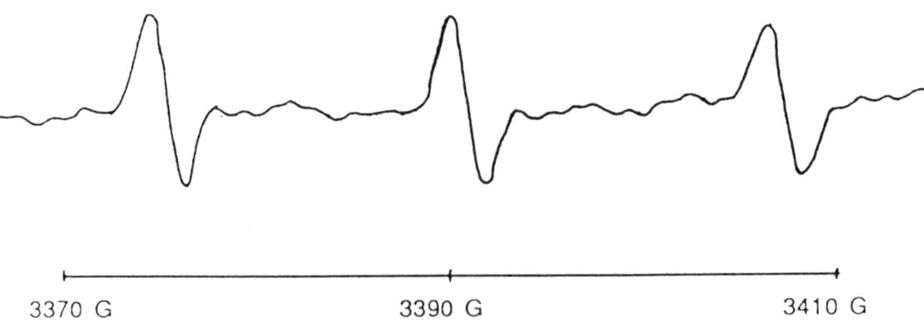

Figure 1. ESR spectra of TAN generated in phosphate buffer (pH 7.55) with 5 µM riboflavin and 20mM TMPD after 10 minutes of illumination.

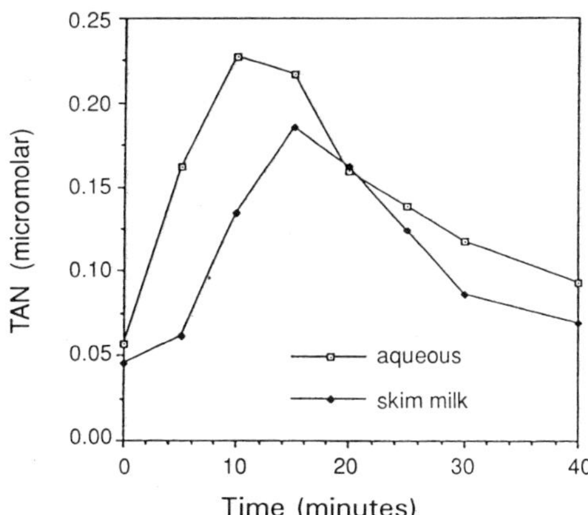

Figure 2. Illumination time *vs* TAN formation in aqueous buffered 5 µM riboflavin and skim milk, both containing 20 mM TMPD.

treatment with Florisil[12,] where added (20 mM) TMPD showed no ESR signal after 30–60 minutes of illumination. This strongly suggests that the interaction of light and riboflavin with TAN in milk is directly linked to the signal decay.

After 15 minutes of bubbling nitrogen into buffered riboflavin/TMPD, then Illuminated in a stoppered flat cell for 5 minutes, where TA ESR spectrum was found. That is, without oxygen present for interaction with excited state riboflavin, the singlet oxygen Type II reaction pathway is prevented. Although superoxide anion generation in milk has been reported, we found no SOD effect on TAN generation, suggesting that superoxide anion was not produced, nor did TMPD react with superoxide anion to form TAN[8]. In conclusion, it was confirmed that singlet oxygen was generated in milk by riboflavin photosensitized oxidation. An earlier report of some aspects of this work has appeared[13].

REFERENCES

1. Foote, C. S. (1968) Photosensitized oxygenation and the role of singlet oxygen. *Acc. Chem. Res.* 1:104–110.
2. Foote, C. S. (1979) Quenching of singlet oxygen. In: H. H. Wasserman, R. W. Murray (Eds.) *Singlet Oxygen,* pp 139–171. Academic Press, New York.
3. McBean, L. D.; Speckman, E. W. (1988) Nutritive value of dairy foods. In: N. P. Wong (Ed.) *Fundamentals of Dairy Chemistry* , pp 343–407. Van Nostrand Reinhold Company: New York
4. Penzer, G. R. (1970) The chemistry of flavins and flavoproteins: aerobic photochemistry. *Biochem. J.* 116:733–743.
5. Grodowski, M. S.; Veyret, B.; Weiss, K. (1977) Photochemistry of flavins. II. Photophysical properties of alloxazines and isoalloxazines. *Photochem. Photobiol.* 26: 341–352.
6. Fritz, B. J.; Matsui, K.; Kasai, S.; Yoshimura, A. (1987) Triplet lifetimes of some flavins. *Photochem. Photobiol.* 45:539–541.
7. Korycka-Dahl, M. B.; Richardson, T. (1978) Activated oxygen species and oxidation of food constituents. *CRC Critical Reviews in Food Science and Nutrition*, p. 209.
8. Aurand, L. W.; Boone, N. H.; Giddings, G. G. (1977) Superoxide and singlet oxygen in milk lipid peroxidation. *J. Dairy Sci.* 60:363–369.
9. Lion, Y.; Delmelle, M.; VanDeVorst, A. (1976) New method of detecting singlet oxygen production. *Nature* 263:442–443.
10. Moan, J., Wold, E. (1979) Detection of singlet oxygen production by ESR, *Nature* 279:450.
11. Chan, T. W., Bruice, T. C. (1977) Reaction of nitroxides with 1,5-dihydroflavins and N,3,5-dimethyl-1,5-dihydrolumiflavin. *J. Am. Chem. Soc.* 99:7287–7291.

12. Aurand, L. W.; Singleton, J. A.; Noble, B. W. (1966) Photooxidation reactions in milk. *J. Dairy Sci.* 49:138–143.
13. L. J. Berliner, D. G. Bradley, D. Meinholtz, D. B. Min, and T. Ogata (1994) The detection of riboflavin photosensitized singlet oxygen formation in milk by electron spin resonance. In: G. Pifat (ed.), Supramolecular Structure and Function, pp. 67–77. Balaban Publishers: Rehovot, Israel.

12

POTENTIAL ANTIOXIDANTS FROM MINOR DIETARY CONSTITUENTS

M. V. Ramana Kumari

Sky Food Co. Ltd.
1-10-8 Nozato
Nishiyodogawaku, Osaka 555, Japan

1. INTRODUCTION

Spices and condiments that are widely used in garnishing and flavouring the food have long been considered as anutrients. Many recent studies demonstrated the potential of some such minor dietarty constituents as anticarcinogens and antioxidants. Natural plant products containing phenols, indoles, aromatic isothiocyanates, methylated flavones, coumarins, sterols, tocopherols, retinol and carotenes are shown to have inhibitory effect on carcinogenesis[1-7]. Apart from exhibiting a direct free radical scavenging, many antioxidants can offer protection to a system via influencing the Phase I and Phase II pathways of the biotransformation process. Therefore, inducing effects on these enzymes are often used to detect the presence of blocking agents in complex natural products. In addition, those compounds that can enrich the endogenous antioxidant levels were also shown to inhibit the process of carcinogenesis[8-9]. Thus research on identifying these natural inhibitors present in food is gaining importance as a key factor in cancer control and prevention. Using the endogenous antioxidant levels and the inducibility of Phase I and Phase II enzymes as parameters, we discussed elsewhere the potential role of mace and clove in chemoprevention [10-12]. The present report, using similar basis, explores the importance of mustard (Brassica nigra) plant products as potential chemopreventive agents.

2. MATERIALS AND METHODS

2.1. Animals

Random bred male Swiss albino mice (7–8 weeks old; body weight 25–30g) maintained in the Animal Facility of Jawaharlal Nehru University were used in the experiments. The animals were provided with standard mouse feed (Hindusthan Lever Ltd., India) and tap water ad libitum, unless otherwise stated.

2.2. Chemicals

Benzo(a)pyrene [B(a)P], bovine serum albumin (BSA), 1-chloro-2–4-dinitrobenzene (CDNB), 2,6-dichlorophenolindophenol (DCPIP), 3,3′-dithiobis 2-nitrobenzoic acid (DTNB), flavin adenine dinucleotide (FAD), glutathione (GSH), 3-hydroxybenzo(a)pyrene [3-HOB(a)P], reduced nicotinamide adenine dinucleotide (NADH), reduced nicotinamide adenine dinucleotide phosphate (NADPH), 2-thiobarbituric acid (TBA), 5-sulfosalicylic acid, and sodium dithionate were obtained from Sigma chemical co., USA. Coomassie Brilliant Blue (G-250) and quinine sulfate were procured from Merck, Germany. Quality grade mustard seeds and oil were purchased from the local market (New Delhi, India).

2.3. Experimental Design

In each experiment, the animals, randomized into three groups (1, 2 and 3) were under treatment for 10, 20 and 30 days respectively. Each group was further assorted into four sub groups (A, B, C and D). In the experiment with mustard seed, subgroup A (control) received normal standard diet, while those of B, C, and D were maintained on 0.5%, 1%, 2% (w/w) mustard seed powder containing diets respectively. For the experiment with mustard oil, subgroup A (control) received distilled water, while animals of B, C, and D were orally administered with 1ml, 2 ml or 4 ml, (kg/bw) of mustard oil respectively. Animals were starved overnight prior to sacrifice.

2.4. Procedures

The animals were sacrificed by cervical dislocation and a 10% (w/v) hepatic homogenate was made in chilled buffer containing 0.154 M KCl and 50 mM Tris-HCl (pH 7.4). A part of it was kept aside for the assay of acid-soluble sulfhydryl (SH) groups and radiation induced malondialdehyde (MDA) formation. The remaining homogenate was subjected to centrifugation at 10,000 X g for 20 min.. The resultant supernatant was further centrifuged (Model L5 50B, Ti- 50 rotor Beckman instruments, Palo Alto, CA) at 100,000 X g for 1h. The supernatant was used for the assay of cytosolic total GST and DTD enzyme activities, while the pellet (microsomal fraction), after resuspension in the homogenizing buffer, was used to assay Cyt. P-450 and Cyt.b$_5$ contents as well as AHH activity. The entire isolation procedure was carried out at 0–4°C. The parameters were assayed in aliquots containing 0.5–1 mg protein/ml.

The contents of Cyt.b$_5$ and Cyt. P-450 were determined spectrophotometrically by the method of Omura and Sato[13] with 185 mM-1cm-1 and 91 mM-1cm-1 as extinction coefficients respectively. The AHH assay was performed fluorimetrically by the procedure of Nebert and Gelboin[14] with B(a)P as substrate. The spectrophotometric estimation of the cytosolic total GST activity was done as described by Habig et al[15]., with CDNB and GSH as substrates and 9.6 mM-1 Cm-1 as the extinction coefficient value. The cytosolic DTD activity was assayed by Benson et al[7]. method using an extinction coefficient value of 6.27

mM-1 cm-1. The acid soluble SH content was estimated with DTNB as reagent, in the supernatants of homogenates precipitated with 5-sulfosalicylic acid as described earlier[10–11].

The hepatic homogenates were exposed to Gamma-radiation (from a ^{60}Co source at a dose rate of 1.3 Gy/ s) for 8 min. and the levels of MDA formed in these samples were estimated as previously reported[12]. The protein content in various subfractions was determined by Bradford's method[16] using BSA as standard. The significance between treated values and respective controls, was evaluated from the Student's t-test.

3. RESULTS AND DISCUSSION

The results obtained by the administration of mustard seed diets and mustard oil are shown in the Tables 1 and 2 respectively. In both the experiments, mustard seed diet and mustard oil administration significantly enhanced SH levels and appreciably decreased radiation-induced MDA formation as compared to their respective controls. While neither mustard seed diet nor mustard oil administration had any significant effect on the AHH activity, elevation in the GST, DTD activities and in Cyt. b_5 and Cyt P-450 contents was observed in the mustard seed diet/ mustard oil treated groups. The elevation, in some cases, was more significant at higher doses/ longer duration levels.

Though there is no single mechanism that can satisfactorily explain the process of chemoprotection, an induction in the electrophile-processing Phase II (detoxicating) enzymes is believed to be responsible for the anticarcinogenic effects of several xenobiotics including antioxidants and other minor dietary constituents of plant origin. Monitoring of Phase II enzyme induction has in fact led to the identification of many new anticarcinogens[2–7]. Since the induction of Phase II system, at times, seems to be dependent on the induction of Phase I system (that can activate carcinogens into their ultimate electrophiles), it is important to study the overall effects on these two systems. The simultaneous enhancement in Phase I and Phase II enzymes indicates a 'bifunctional'[17] inducer-like behaviour of mustard seed and mustard oil. Results show an elevation in the SH levels in concomitance with enhanced GST levels and a reduction in radiation induced lipid peroxidation, demonstrating an enrichment in the endogenous antioxidant defense pool. The data also suggests that these positive modulatory influences observed here might contribute as additional factors (other than the direct free radical scavenging) in the anticarcinogenicity of mustard seed and mustard oil. Also, the anticarcinogenic properties of other cruciferous vegetables are being attributed to the indole glucosinates and isothiocyanates[18]. These compounds might, at least partly, be responsible for the effects reported here. In view of the fact that various mustard products are routinely consumed, taken together with the hypothesis that carcinogenesis can be inhibited by the modulation of metabolic enzymes, it is worthwhile to further elucidate the potential of these products as chemopreventive agents.

4. ACKNOWLEDGMENTS

This work was done when the author was at the School of Life Sciences, Jawaharlal Nehru University, New Delhi, India and the grant was provided by the Indian Council of Medical Research, New Delhi, India.

Table 1. Effects of mustard seed administration on mouse liver

Group Details	Treatment Details	GST (μmols CDNB conjugate formed/min/ mg. protein)	DTD (nmols /mg. protein)	Cyt.b_5 (nmols / mg. protein)	Cyt. P-450 (nmols / mg protein)	AHH (pmols 3OHB(a)P formed/min/ mg protein)	Protein content (mg/ g tissue) Microsomal	Protein content (mg/ g tissue) Cytosolic	Acid soluble SH level (μmols/g tissue)	Lipid peroxide (nmols MDA formed/ mg protein)
Group 1	10 DAYS									
A	Control	6.42±0.15	6.53±0.17	0.08±0.01	0.79±0.02	400±26	10.53±0.90	7.89±0.62	14.01±0.67	2.77±0.08
B	Mustard seed (0.5%)	10.03±0.35[a]	9.03±0.50[a]	0.13±0.01[a]	0.87±0.04[e]	420±20	11.05±0.87	8.21±0.53	26.44±0.40[a]	2.62±0.15
C	Mustardseed (1%)	10.31±0.40[a]	9.81±0.42[a]	0.15±0.02[a]	0.89±0.06[e]	425±15	11.12±0.77	8.08±0.50	26.51±0.71[a]	2.50±0.06
D	Mustard seed (2%)	11.14±0.36[a]	11.65±0.27[a]	0.18±0.02[a]	0.90±0.03[e]	422±13	11.13±0.80	8.09±0.41	26.73±0.15[a]	2.56±0.22
Group 2	20 Days									
A	Control	6.44±0.20	7.40±0.20	0.07±0.05	0.78±0.06	410±15	10.81±0.70	7.91±0.57	15.77±0.57	2.71±0.15
B	Mustard seed (0.5%)	11.38±0.41[a]	9.78±0.31[a]	0.13±0.01[a]	0.90±0.03[c]	411±12	10.93±0.52	8.15±0.68	27.31±0.82[a]	2.46±0.06[f]
C	Mustardseed (1%)	11.87±0.62[a]	9.89±0.30[a]	0.16±0.02[a]	0.93±0.02[d]	422±15	10.97±0.60	8.17±0.71	27.73±0.51[a]	2.35±0.06[e]
D	Mustard seed (2%)	12.88±0.78[a]	11.95±0.42[a]	0.19±0.03[a]	0.95±0.01[c]	436±19	10.99±0.50	8.16±0.60	27.99±0.13[a]	2.01±0.19[e]
Group 3	30 Days									
A	Control	6.51±0.18	7.79±0.25	0.07±0.01	0.78±0.05	411±16	10.89±0.53	8.05±0.62	19.10±0.23	2.78±0.22
B	Mustard seed (0.5%)	13.54±1.01[a]	12.03±0.07[a]	0.15±0.02[a]	0.90±0.02[e]	442±19	11.13±0.80	8.18±0.70	30.29±0.13[a]	2.33±0.03[f]
C	Mustard seed (1%)	15.96±0.08[a]	12.02±0.01[a]	0.16±0.03[a]	0.96±0.01[c]	443±17	11.08±0.35	8.16±0.31	35.52±0.52[a]	2.07±0.07[e]
D	Mustard seed (2%)	16.91±0..83[a]	12.03±0.01[a]	0.32±0.04[a]	0.99±0.01[c]	440±17	11.38±0.40	8.20±0.42	38.98±0.24[a]	1.17±0.01[a]

Values Expressed As Mean± S.E of 8 - 12 Animals

Superscripted values Significantly Differ From Their Respective Control Values-

a : P< 0.0005; b : P < 0.001; c : P< 0.0025; d : P < 0.005 ; e : P< 0.01; f : P < 0.025

Table 2. Effects of mustard oil administration on mouse liver

Group Details	Treatment Details	GST (µmols CDNB conjugate formed/min/ mg. protein)	DTD (nmols /mg. protein)	Cyt.b_5 (nmols / mg. protein)	Cyt. P-450 (nmols / mg protein)	AHH (pmols 3OHB(a)P formed/min/ mg protein)	Protein content (mg/ g tissue) Microsomal	Protein content (mg/ g tissue) Cytosolic	Acid soluble SH level (µmols/g tissue)	Lipid peroxide (nmols MDA formed/ mg protein)
Group 1	**10 DAYS**									
A	Control	7.64±0.13	6.66±0.13	0.06±0.02	0.80±0.02	420±10	10.01±0.89	6.82±0.03	15.66±1.23	2.52±0.52
B	Mustard oil (1ml/kg bw)	7.82±0.22	8.71±0.30[a]	0.12±0.02[a]	0.81±0.06	415±19	10.21±0.37	7.92±0.61	20.12±1.03[b]	1.08±0.34[a]
C	Mustard oil (2ml/kg bw)	7.86±0.30	8.80±0.33[a]	0.14±0.01[a]	0.82±0.03	415±18	11.03±0.91	8.89±0.71[f]	22.37±0.98[a]	1.03±0.33[a]
D	Mustard oil (4ml/kg bw)	8.96±0.41[f]	8.99±0.71[a]	0.14±0.02[a]	0.81±0.03	419±11	12.10±0.72[d]	9.93±0.51[d]	26.34±1.14[a]	1.01±0.02[a]
Group 2	**20 Days**									
A	Control	7.82±0.12	6.71±0.42	0.08±0.02	0.82±0.03	430±20	10.81±0.77	6.93±0.07	17.89±0.99	2.62±0.71
B	Mustard oil (1ml/kg bw)	9.01±0.16[e]	8.93±0.56[a]	0.13±0.05[a]	0.86±0.02	434±16	10.92±0.65	7.99±0.81	22.89±0.89[b]	1.06±0.62[a]
C	Mustard oil (2ml/kg bw)	9.37±0.32[e]	9.95±0.71[a]	0.14±0.01[a]	0.87±0.04	436±15	11.01±0.50	8.01±0.70[f]	27.78±1.19[a]	1.00±0.31[a]
D	Mustard oil (4ml/kg bw)	9.99±0.07[d]	10.01±0.81[a]	0.14±0.03[a]	0.87±0.01	440±10	12.04±0.52[d]	10.03±0.92[a]	30.01±1.17[a]	1.01±0.20[a]
Group 3	**30 Days**									
A	Control	8.01±0.06	6.81±0.30	0.08±0.02	0.81±0.03	435±25	10.92±0.91	7.01±0.15	20.15±1.20	2.77±0.66
B	Mustard oil (1ml/kg bw)	10.15±0.72[f]	10.03±0.42[a]	0.13±0.06[a]	0.92±0.04[e]	440±18	11.90±0.87	7.89±0.37	25.21±1.40[b]	1.00±0.05[a]
C	Mustard oil (2ml/kg bw)	11.32±0.81[e]	11.13±0.56[a]	0.16±0.08[a]	0.93±0.03[e]	445±19	12.01±0.57[f]	9.91±0.61[d]	32.44±1.32[a]	0.98±0.30[a]
D	Mustard oil (4ml/kg bw)	12.15±0.09[a]	11.15±0.40[a]	0.18±0.10[e]	0.93±0.04[e]	450±30	12.03±0.60[f]	12.34±0.05[a]	33.56±1.07[a]	0.87±0.41[a]

Values Expressed As Mean± S.E of 8 - 12 Animals

Superscripted values Significantly Differ From Their Respective Control Values-

a : P< 0.0005; b : P < 0.001; c : P< 0.0025; d : P < 0.005 ; e : P< 0.01; f : P < 0.025

REFERENCES

1. Newmark, H.L. (1987) Plant phenolics as inhibitors of mutational and precarcinogenic events. Can. J. Physiol. Pharmacol. 65: 461–466.
2. Sparnins, V. L., Venegas, P. L. and Wattenberg, L. W. (1982) Glutathione S- transferase activity by compounds inhibiting chemical carcinogenesis and by dietary constituents. J. Natl. Cancer Inst. 68,: 493–496.
3. Wattenberg, L.W. (1983) Inhibition of neoplasia by minor dietary constituents. Cancer Res. (Suppl.). 43: 2448s-2453s.
4. Wattenberg, L.W. (1985) Chemoprevention of cancer. Cancer Res. 45: 1–8. 5. Williams, G.M. (1985) Modulation of chemical carcinogenesis by xenobiotics. Fundam. Appl. Toxicol. 4: 325–344.
5. Williams, G. M. (1985) Modulation of chemical carcinogenesis by xenobiotics. Fundam. Appl. Toxicol. 4: 325–344.
6. Benson, A. M., Cha, Y. N., Bueding, E., Heine, H. S. and Talalay, P. (1979) Evaluation of extrahepatic glutathione S- transferase activities and epoxide hydratase activities by 2(3)- tert-butyl-4- hydroxyanisole. Cancer Res. 39: 2971–2977.
7. Benson, A. M., Heukeler, M. J. and Talalay, P. (1980) Increase of NAD(P)H: Quinone reductase by dietary antioxidants: possible role in protection against carcinogenesis and toxicity. Proc. Natl. Acad. Sci. U. S. A. 77: 5216–5220.
8. Horton, A. A. and Fairhurst, S. (1987) Lipid peroxidation and mechanisms of toxicity. Crit. Rev. Toxicol. 18: 27–79.
9. Puglia, C.D. and Powell, S.R. (1984) Inhibition of cellular antioxidants: A possible mechanism of toxic cell injury. Environ. Heal. Persp. 57: 307–311.
10. Kumari, M.V.R. and Rao, A.R. (1989) Effects of mace (Myristica fragrans Houtt.) on cytosolic glutathione S-transferase activity and acid soluble sulfhydryl level in mouse liver. Cancer Lett. 46: 87–91.
11. Kumari, M.V.R. (1991) Modulatory influences of clove (Caryophyllus aromaticus, L) on hepatic detoxification systems and bone marrow genotoxicity in male Swiss albino mice. Cancer Lett. 60: 67–73.
12. Kumari, M.V.R. (1992) Modulatory influences of mace (Myristica fragrans, Houtt.) on hepatic detoxification systems and bone marrow genotoxicity in male Swiss albino mice. Nutr. Res. 12: 385- 394.
13. Omura, T. and Sato, R. (1964) The cabon monoxide binding pigment of liver microsomes. J. Biol. Chem. 239: 2370- 2378.
14. Nebert, D. W. and Gelboin, H. V. (1968) Substrate -inducible microsomal aryl hydroxylase in mammalian cell culture I. Assay and properties of induced enzyme. J. Biol. Chem. 243: 6242- 6249.
15. Habig, W. H., Pabst, M. J. and Jakoby, W. H. (1974) Glutathione S- transferases. The first enzymatic step in mercapturic acid formation. J. Biol. Chem. 249: 7130–713.
16. Bradford, M.M. (1976) A rapid and sensitive method for the quantitation of microgram quantities of protein utilizing the principle of protein-dye binding. Anal. Biochem. 72: 248–254.
17. Prochaska, H.J. and Talalay, P. (1988) Regulatory mechanisms of monofunctional and bifunctional anticarcinogenic enzyme inducers in murine liver. Cancer Res. 48: 4776- 4782.
18. McDanell, R., McLean, A. E., Hanley, A. B. and Fenwick, G.R. (1988) . Chemical and biological properties of indole glucosinates (glucobrassicins): a review. Food and Chem.Toxicol. 26: 59–70.

13

ANTIOXIDATIVE EFFECTS OF GINKGO BILOBA EXTRACT (GBE)

Takashi Miyajima, Toshikazu Yoshikawa, and Motoharu Kondo

First Department of Medicine
Kyoto Prefectural University of Medicine
Kyoto 605, Japan

1. INTRODUCTION

Recently, various kind of anti-oxidative effects of natural foods and products have been reported. GINKGO BILOBA EXTRACT (GBE) is an extract from the leaves of Ginkgo (maiden tree). GBE is clinically used in many European countries for the therapy of cerebro-vascular diseases and senile dementia. And it is regarded that free radicals play a role in the pathogenesis of these diseases. In this study, we investigated the anti-oxidative properties of GBE.

2. MATERIALS AND METHOD

GBE was a kind gift from Sunwell Co. Ltd. (Tokyo, Japan).

2.1. The Free Radicals

We applied electron spin resonance spectroscopy (JES FR80, JOEL, Tokyo, Japan) to investigate the free radicals scavenging activities. We measured the GBE's scavenging activities of superoxide anion radicals, hydroxyl radicals, and diphenyl-p-picryihydrazyl (DPPH) radicals. Superoxide anion radicals were generated by xanthine-xanthine oxidase system. Hydroxyl radicals were generated by Fenton's reaction. As a spin trapping agent, 5, 5-dimethyl-1-pyroline-N-oxide (DMPO) was used.

2.2. The Effects of GBE on Lipid-Peroxidation

We examined the effects of GBE on auto-oxidation (lipid-peroxide) of rat brain homogenate both in vitro and ex vivo.

Food and Free Radicals, edited by Hiramatsu *et al.*
Plenum Press, New York, 1997

In vitro: A mixture of 5% rat brain homogenate (0.01 M PBS: pH7.4) and GBE suspension (concentration: 0, 3, 10, 30, 100 mg/ml) was incubated at 37°C for 0, 30, 60, 180 minutes under the air. After the incubation, thiobarbituric acid-reactive substances (TBA-RS) was measured by Ohkawa's method.

Ex vivo: GBE (10mg/kg.rat) was orally administered one hour before sacrifice. 10% rat brain homogenate in 0.01M PBS was incubated at 37°C for 0, 30, 60, 180 minutes under the air. After the incubation, TBA-RS was measured.

3. RESULTS

3.1 . The Free Radicals Scavenging Effects

GBE scavenged superoxide anion radicals dose-dependently. IC_{50} to superoxide anion was about 20 mg/ml. GBE also scavenged DPPH radicals in a dose dependent manner. IC_{50} to DPPH radicals was about 30 mg/ml. Scavenging activity of GBE to hydroxyl radicals was not so strong compared to that superoxide and DPPH radicals.

3.2. The Effects of GBE on Lipid-Peroxidation

GBE inhibited the lipid-peroxidation of rat brain homogenate both *in vitro* and *ex vivo*. In in vitro study, GBE inhibited the lipid-peroxidation dose-dependently and the lipid-peroxidation was significantly inhibit at 30 mg/ml.

4. CONCLUSION

GBE is clinically administrated for the treatment of cerebral dysfunction, such as; difficulties of memory, dizziness, tinnutis, headache, and emotional instability with anxiety[1]. These actions is due to GBE's vasoregulating activities (increased blood flow), platelet activating factor antagonism, metabolic changes (changes in neuron metabolism and a beneficial influence on neurotransmitter disturbances), and prevention of membrane damage caused by free radicals.

In this study, we proved the anti-oxidant activities of GBE (free radical scavenging effects using ESR spectroscopy and effects on lipid-peroxidation of brain homogenete). There are other reports that suppost present results that GBE has a superoxide anion radical scavenging activity[2,3]. Flavonoids are known to scavenge superoxide anion radicals[4]. GBE contains flavonoids (ginkgo-flavone glycoside) and terpenoids (ginkgolides and bilobalide). The antioxidant effects of GBE may be due to flavonoids.

From the results of this study, we conclude that GBE possesses free radical scavenging activities and these activities may contribute the improvement of cerebral dysfunction induced by free radicals (oxidative stress).

5. REFERENCES

1. Kleijnen, J., Knipschild, P. (1992) Lancet, Nov. 340 (8828); 1136–1139.
2. Pincemail, J., Thirion, A., Dupus, M., et al. (1987) Experientia. 43; 181–184.
3. Pincemail, J., Dupus, M., Nasr, C., et al. (1989) Experientia. 45; 708–712.
4. Robak, J., Gryglewski, R.J. (1989) Biochem Pharmacol. 37;837–841.

14

SPIRULINA COMPOUND

Minoru Yoneda[1] and Akio Fujikawa[2]

[1]Aoba Spirulina Co., Ltd,
325 Sashimono-cho, Ebisugawa-agaru
Kawaramachi-dori, Nakagyo-ku, Kyoto
[2]Food and Cosmetic Science Laboratory
2-7 Kaidokuchi-cho, Fukakusa, Fushimi-ku, Kyoto, Japan

1. INTRODUCTION

Spirulina, a kind of blue alga, has been long used as a food-stuff because it is rich in the content of well-balanced nutrients. As it has very potent photo-synthetic activities, the antioxidative capacity has also been regarded extremely high. However, when actually ingested, the advantage of antioxidative potential is erratic by individuals, and not necessarily reproduced clearly. To enhance the functional effects of spirulina reproducibly, we have prepared a spirulina compound by adding small amounts of several kinds of food materials to spirulina, and studied its effects on alopecia and liver spots (chloasma) which are symptoms most likely to be caused by free radicals.

2. MATERIALS AND METHODS

The Spirulina compound was prepared by adding small amounts (0.5–5% each) of Aspalathus linearis, acerola, yeast, milk calcium, dried lactobacillus and other powders and extracts to Spirulina maxima, the main ingredient, and blending them. It was invented by Aoba Spirulina Co., Ltd (Trial product name: Spirulina GAEA Ace).

2.1 Effects on Experimental Alopecia

Four rats of the same specification were used: two rats in Group A were fed for two weeks with usual rat pellets, and two in Group B for the same period with the same pellets that contain the Spirulina compound substituting 1/6 weighted portion of the pellets. Thereafter, the hair on the dorsal side of the rats in Groups A and B was clipped, where linoleic acid ester was applied that had been subjected to UV irradiation for 24h under agitation with air. The ester was washed away after 10 min. The same procedures were

Table 1.

| Group | Age | TBA (mg/20 cm^2) | |
		Pre-dose	Post-dose
A			
	37	0.03	0.03
	47	0.04	0.04
	46	0.02	0.03
	Average	0.03	0.033
B			
	43	0.02	0.01
	47	0.06	0.02
	41	0.03	0.01
	Average	0.036	0.013

repeated next day. Remarkable alopecia occurred in Group A at 3–4 days thereafter, while in Group B appeared only slight alopecia in one rat, and no alopecia in another one.

2.2 Effects on Liver Spots

Six females with obvious liver spots on the face were divided into two Groups A and B. Usual meals were provided to three females in Group A for three months, while usual meals plus 6g/day of the Spirulina compound to three females in Group B for the same period. Thus, the pre-dose and post-dose lipid peroxide values (TBA values) in the facial sebum were compared. As shown in Table 1, an improvement was apparent in the lipid peroxide values in the facial sebum for the latter group. The facial sebum was sampled in accordance with Hayakawa's method, and determined by the TBA method.

3. SUMMARY

Alopecia and liver spots are most stressful troubles in men and women respectively, but as they are not likely to advance into serious diseases, they are apt to remain untreated, with countermeasures as well as causes not clearly defined. Nevertheless, it appears that alopecia and liver spots are worsened as lipid peroxide is generated and actually its remarkable increase is observed in the serum and sebum of those suffering from the troubles. It is very significant that the Spirulina compound has shown inhibitory activities against lipid peroxide with good reproducibility, in contrast with not necessarily explicit effects of spirulina when used as it is. The Spirulina compound will be the product that can be expected to reduce cosmetic problems and thereby drive out stresses.

15

ANTIOXIDANT BEVERAGE "βCATECHIN"

Tadashi Yoneda,[1] Midori Hiramatsu,[2] Michiko Sakamoto,[1] Keiichi Togasaki,[1] Makiko Komatsu,[2] and Kiyomichi Yamaguchi[3]

[1]Research and Development Department
Shimakyu Chemical Co., Ltd., Osaka
[2]Division of Medical Science, Institute for Life Support Technology
Yamagata Technopolis Foundation, Yamagata
[3]Sky Food Co., Ltd.
Osaka, Japan

1. INTRODUCTION

"βCATECHIN" (SKY FOOD Co., Ltd., Osaka) was developed with the intention of producing an antioxidant beverage. The production depends on a green tea extract as the main component. Water-soluble ascorbic acid, sunflower seed extract, fat-soluble dunaliella carotene and natural vitamin E are added during production. We examined the effect of "βCATECHIN" on free radical scavenging and on the thiobarbituric acid reactive substances (TBARS) formation, and superoxide dismutase (SOD) activity in the cerebral cortex of $FeCl_3$-induced epileptogenic rats.

2. MATERIAL AND METHOD

2.1 Scavenging Activities for Free Radicals

Free radical scavenging activities were analyzed by an electron spin resonance (ESR) spectrometry.

2.2 Iron-Induced Epileptogenic Focus of Rats

Male adult Wistar rats weighing 220–250g were used. Five microliters of 100 mM $FeCl_3$ solution were injected into the left sensory motor cortex of rat according to the method of Willmore et al.[1]. "βCATECHIN" was administered orally to rats 30 min before injection of the $FeCl_3$ solution. Each rat was killed 30 min after iron-solution injection, the brain was removed, and the cortex was dissected. TBARS were analyzed by the method of Ohkawa et al.[2] and the SOD activity was analyzed by the method of Hiramatsu et al.[3].

Food and Free Radicals, edited by Hiramatsu *et al.*
Plenum Press, New York, 1997

Table 1. Effect of "βCATECHIN" on free radicals

Concentration of "βCATECHIN"(%)	Spin number ($\times 10^{14}$)		
	DPPH radicals	Superoxide	Hydroxyl radicals
0	6.1	4.2	3.5
0.01	4.1	3.8	—
0.05	0.0	—	—
0.1	—	2.6	3.7
1	—	0.3	2.0
10	—	0.0	1.5
100	—	—	0.4

3. RESULTS

The free radical scavenging effects of "βCATECHIN" are shown in Table 1. A 0.05% solution of "βCATECHIN" completely scavenged 1,1-diphenyl-2-picrylhydrazyl (DPPH) radicals. A 10 % solution of "βCATECHIN" completely scavenged superoxide as DMPO spin adducts generated by hypoxanthine-xanthine oxidase system. An undiluted solution of "βCATECHIN" scavenged about 90 % of hydroxyl radicals as DMPO spin adducts generated by H_2O_2-$FeSO_4$-DETAPAC system. The pattern of scavenging activities by "βCATECHIN" was dose-dependent for superoxide, DPPH, and hydroxyl radicals.

The effect of "βCATECHIN" on $FeCl_3$-injected isocortex are shown in Table 2. The TBARS level increased significantly following iron-solution injection into the cortex when compared to control. "βCATECHIN" inhibited TBARS formation in a dose-dependent manner. SOD activity was significantly decreased by isocortical iron-solution injection. Following pretreatment with 2 ml/kg body weight of "βCATECHIN" the expected decrease in SOD activity was no longer apparent. Further "βCATECHIN" administration caused significant increase in SOD activity in saline control animals when compared to baseline of SOD activity in uninjected animals.

4. DISCUSSION

In this study "βCATECHIN" had a dose-dependent scavenging effect on free radicals and inhibited TBARS formation in the iron-injected cortex of the rat. These data sug-

Table 2. Effect of "βCATECHIN" on thiobarbituric acid reactive substances (TBARS) level and superoxide dismutase (SOD) activity in the iron-induced epileptic rats

	TBARS (n mol/g tissue)	SOD activity ($\times 10^3$ unit/g tissue)
CS	47.4 ± 4.3	1.44 ± 0.05
CF	174.4 ± 35.4 [a]	1.24 ± 0.04 [a]
B1F	71.9 ± 5.7 [b]	1.20 ± 0.03 [a]
B2F	54.7 ± 3.3 [b]	1.48 ± 0.06
B2S	42.2 ± 2.8	1.62 ± 0.04 [b]

The results were expressed as a means $\pm$ S. E. M. (n=7).
[a] $p<0.005$ versus CS
[b] $p<0.01$ versus CS.
CS: control; CF: $FeCl_3$ solution injection; B1F: administration of "βCATECHIN" (1 ml/kg body weight) and $FeCl_3$ solution injection; B2F: administration of "βCATECHIN" (2 ml/kg body weight) and $FeCl_3$ solution injection; B2S: administration of "βCATECHIN" (2 ml/kg body weight) and saline injection.

gest that the inhibition of cortical iron-induced TBARS formation by "βCATECHIN" may be the result of the scavenging action of "βCATECHIN" on iron generated free radicals. These observations imply that active scavenging compounds may pass the blood brain barrier.

Terao et al. [4] have suggested that flavonoids can serve as scavenger of aqueous peroxyl radicals near membrane surfaces. "βCATECHIN" contains green tea extract as a radical scavenger with membrane amphiphillic, ascorbic acid and sunflower seed extract as water-soluble radical scavengers, dunaliella carotene and natural vitamin E as fat-soluble radical scavengers. Ascorbic acid can regenerate tocopherol resulting in a synergistic effects [5]. A synergistic effect with tocopherol and ascorbic acid has been demonstrated by (-)-epigallocatechin gallate [6].

Thus, the compounds in "βCATECHIN" may be effective free radical scavenger and inhibit lipid peroxidation by a synergistic effect.

In conclusion, we suggest that "βCATECHIN" may be an antioxidant beverage with prophylactic effects on neurological diseases with mechanism of pathogenesis related to free radical systems.

REFERENCES

1. Willmore, L. J. ; Sypert, G. W. ; Munson, J. B. (1978) Science 200 : 1501–1503.
2. Ohkawa, H. ; Ohnishi, N. ; Yagi, K. (1979) Anal. Biochem. 95 : 351–358.
3. Hiramatsu, M. ; Kohno, M. (1986) JOEL News 26 : 106–109.
4. Terao, J. ; Piskula, M. ; Yao, O. (1994) Arch. Biochem. Biophys. 308 : 278–284.
5. Niki, E. ; Kawakami, A. ; Yamamoto, Y. ; Kamiya, Y. (1985) Bull. Chem. Soc. Jpn. 58 : 1971- 1975.
6. Matsuzaki, T. ; Hara, Y. (1985) J. Agr. Chem. Soc. Jap. 59 : 129–134.

16

OXIDATIVE STRESS AND ANTIOXIDANT BIOFACTOR (AOB®)

Yukiko Minamiyama,[1] Eisuke Sato,[1] Shigekazu Takemura,[1] Masayasu Inoue,[1] and Toshikazu Yoshikawa[2]

[1]Department of Biochemistry
Osaka City University Medical School
1-4-54 Asahimachi, Abeno-ku, Osaka 545, Japan
[2]Department of Medicine
Kyoto Prefectural University of Medicine
Kawaramachi-hirokoji, Kamikyo-ku, Kyoto 602, Japan

1. INTRODUCTION

Antioxidant biofactor (AOB®) is an unique processed grain food obtained from soybean, wheat, rice germ, rice bran, tear grass, green tea, sesame, citron, green leaves extracts and malted rice. These ingredients were degraded to low molecular weight compounds by roast and fermentation with *Aspergilus oryzae* (Fig. 1). AOB contains a variety of compounds with antioxidant activity including flavonoids, α-tocopherol, vitamin C, tannins and vitamin B_2. These ingredients have potent activity to scavenge reactive oxygen species, such as superoxide radical dose-dependently. Plasma of AOB-treated rats also scavenge superoxide and diphenyl-p-picrylhydrazyl (DPPH) radicals[1]. Fenton's reaction and lipid peroxidation are also inhibited by AOB[1]. Thus, AOB is a potent antioxidant both *in vitro* and *in vivo*. The present work demonstrates its effect on ischemia/reperfusion-induced renal injury and the fate of NO in the rat.

2. MATERIALS AND METHODS

AOB and NOC7, a nitric oxide donor (Fig. 2), were obtained from AOA Japan Co. LTD (Kobe) and Dojin Co. LTD (Kumamoto), respectively. After oral administration of AOB-mixed diet (1, 3, or 5 g AOB + normal diet) (total diets ; 20 g) / rat / day, rats were used for some experiments. After 5 g AOB administration for 7 days, renal injury was elicited in male Sprague-Dawley rats (200 g) by 1 hr occlusion and 24 hr reperfusion of the left renal artery and vein. Control rat was given 20 g normal diet per day. At indicated times, lipid peroxidation products in the kidney and urea nitrogen (BUN) in plasma were determined.

Food and Free Radicals, edited by Hiramatsu *et al.*
Plenum Press, New York, 1997

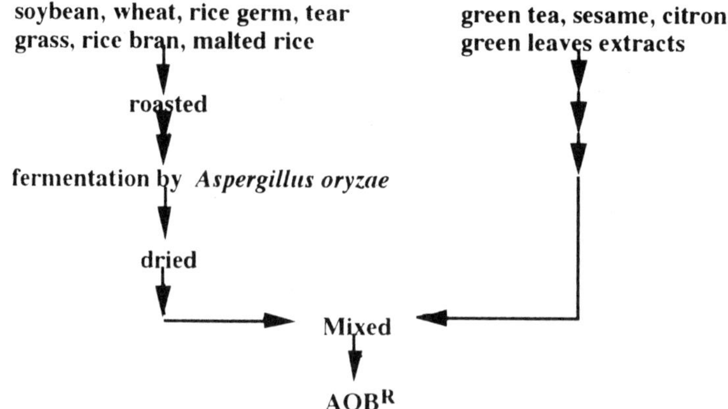

Figure 1. Manufacturing process of AOB.

After administration of either normal diet or 3 g AOB containing diet for 3 days, male Wistar rats (200 g) were intravenously injected with NOC7. The changes in blood pressure and levels of nitrosyl-hemoglobin (NO-Hb) in the circulating blood was measured by manometer and ESR spectrometer, respectively.

Under urethane anesthesia (1 g/kg, i.p.), NOC7 (10 μmol/kg) was intravenously injected to the rats and the change in blood levels of NO-Hb was analyzed by ESR. At indicated times, 400 μl of blood samples were collected from the femoral vein in ESR tubes (inner diameter of 4 mm). Blood samples were quickly frozen in liquid nitrogen and subjected to ESR analysis at 110ûK using a JES-RE1X spectrometer (JEOL, Tokyo) with 100 kHz field modulation. The microwave power was attenuated to 8 mW. ESR analysis was carried out with microwave frequency of 9.108 GHz, 325 ± 250 mT field, 1 min sweep time, 0.63 mT modulation amplitude, and 0.03 s of time constant.

3. RESULTS AND DISCUSSION

AOB inhibited the production of superoxide radical dose-dependently both *in vitro* and in plasma of AOB-treated rats (Fig. 3A)[1]. AOB also scavenged DPPH radicals, and inhibited Fenton reaction and lipid peroxidation as determined with brain homogenate[1]. AOB also inhibited generation of oxygen radicals by opsonized-zymosan-stimulated oral neutrophils (PMN) (Fig. 3B).

Ischemia followed by reperfusion increased BUN and induced renal failure. Administration of AOB significantly inhibited the damages and the increase of BUN induced by post-ischemic reperfusion (Fig. 4).

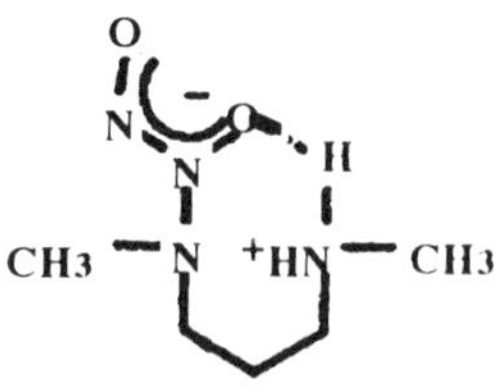

Figure 2. Chemical structure of NOC-7. NOC-7 spontaneously releases NO, without requiring any co-factor.

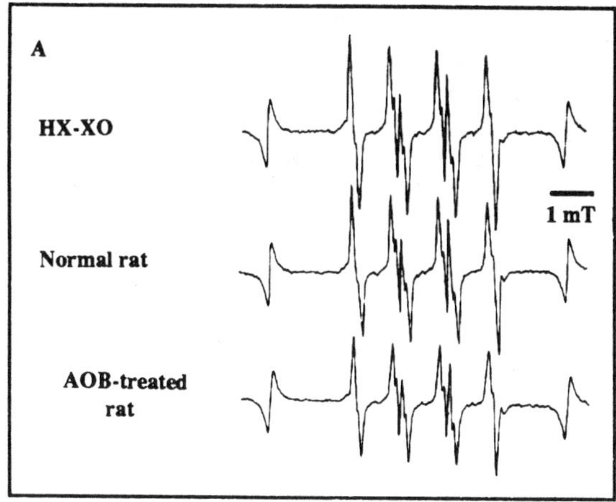

Figure 3. A. O_2^- scavenging activity of plasma for animals fed on normal diet and AOB-containing diet. Superoxide was generated by 0.1 mM hypoxanthine and xanthine oxidase (0.1 U/ml). Data from reference[1]. B. Effects of AOB on the generation of oxygen radicals by opsonized-zymonsan (OZ)-stimulated oral PMN. 5 × 10^5 cells and 100 μM luminol were incubated at 37°C with or without AOB, and stimulated by 20 μg/ml OZ.

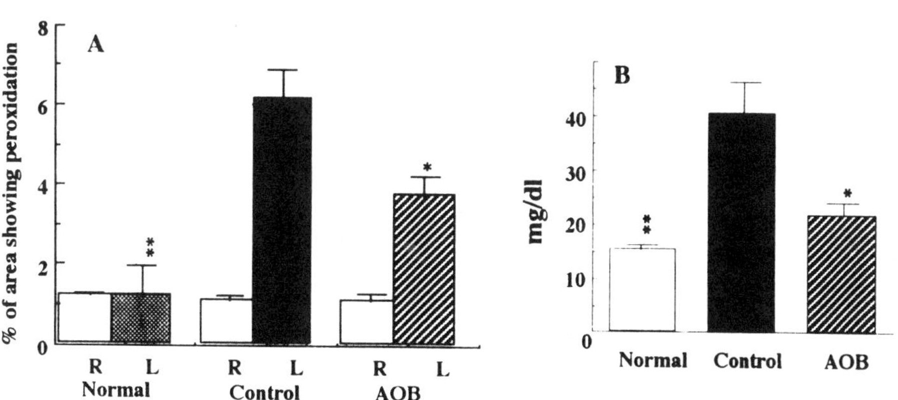

Figure 4. Effects of AOB on renal injury induced by ischemia/reperfusion. A. Histochemical determination of lipid peroxidation in the kidney using image analyzer. B. Changes in levels of blood urea nitrogen (BUN). Values are mean ± SEM. *: P < 0.05; **: p < 0.01 vs. the left control kidney.

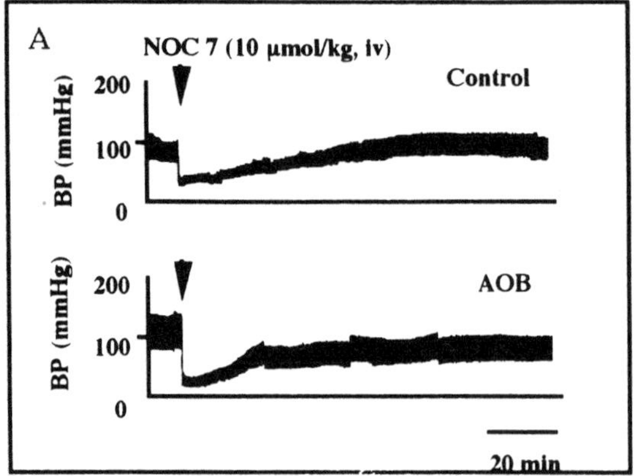

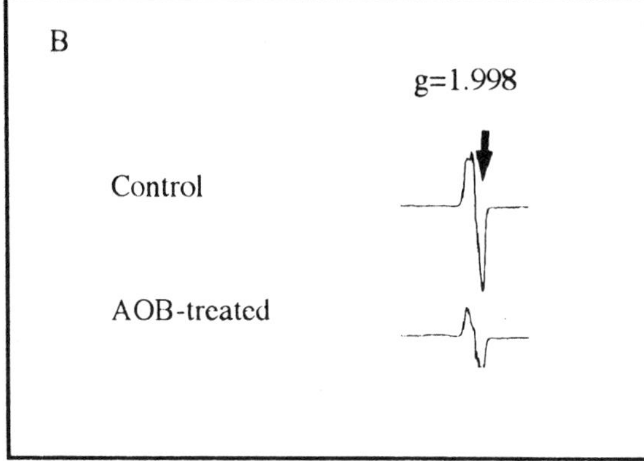

Figure 5. Effect of AOB on the changes in blood pressure (A) and NO-Hb levels (B) induced by NOC7. NO-Hb signals in blood samples were determined at 10 min after NOC7 injection.

The effect of AOB was also studied on the fate of NO generated from NOC7 by measuring the formation of NO-Hb in the circulating RBC. Administration of NOC7 decreased the blood pressure and increased the levels of NO-Hb in the circulation of control rats. AOB enhanced the depressor action of NOC7 but decreased the level of NO-Hb (Fig. 5).

These results suggested that AOB$^{®}$ might protect against renal injury induced by postischemic reperfusion, enhance NO-dependent relaxation of resistant arteries and modulate NO metabolism.

ACKNOWLEDGMENTS

We thank AOA Japan Co. and Dojin Co. for kindly providing us with AOB$^{®}$ and NOC-7.

REFERENCE

1. Minamiyama, Y., Yoshikawa, T., Tanigawa T., Takahashi S., Naito Y., Ichikawa, H., Kondo M.: Antioxidative effects of a processed grain food (1994) J. Nutr. Sci. Vitaminol. 40: 467–477.

17

MANDA SCAVENGES FREE RADICALS AND INHIBITS LIPID PEROXIDATION IN IRON-INDUCED EPILEPTIC FOCUS IN RATS

M. Kawai[1] and S. Matsuura[2]

[1]Department of Neuroscience
Institute of Molecular and Cellular Medicine
Okayama University Medical School
2-5-1 Shikata-cho, Okayama 700, Japan
[2]Manda Fermentation Co. Ltd.
Hiroshima, Japan

1. INTRODUCTION

A number of studies have suggested that ischemia [1], trauma [2], vascular injury [3], cancer [4], aging [5] and neurological disorders such as epilepsy [6], are associated with free radicals and radical-mediated peroxidation reaction. The involvement of free radicals in such disorders opens the possibility of prevention and therapy by the use of many kind of antioxidant, including fermentation products [7] and Chinese medical herbal preparations[8]. Manda is a health food commercially sold in Japan. It is made by yeast fermentation of cane sugar, fruits, seeds, vegetables and seaweeds for more than 39 months [9].

In the present study, we examined the antioxidant effects of Manda on neural lipid peroxidation in an iron-induced epileptic focus in rats, and on reactive oxygen species *in vitro*.

2. MATERIALS AND METHODS

2.1 Chemical and Animals

Manda, a brown, sweet and sticky fermented natural food, was provided by Manda Fermentation Co., Ltd. (Hiroshima, Japan). Hypoxanthine (HPX), 1,1-diphenyl-2- picryl-hydrazyl (DPPH), diethylenetriaminepentaacetic acid (DTPA) were purchased from the Sigma Chemical Co. (St. Louis, MO, USA), xanthine oxidase (XOD) and 5, 5'-dimethyl-

Food and Free Radicals, edited by Hiramatsu *et al.*
Plenum Press, New York, 1997

1-pyrroline-N-oxide (DMPO) were from Boehringer Gmbh (Mannheim, Germany) and LABOTEC, Ltd. (Tokyo, Japan), respectively.

Male Sprague-Dawley rats from Clea Japan, Inc. (Tokyo, Japan) weighing 240–260 g were used in this study and housed at a constant temperature ($25 \pm 2°C$) and a humidity ($50 \pm 5\%$). Control standard diet (MF; Oriental Yeast Co. Ltd., Japan) and water were provided freely for 7 days until the experiment.

2.2 Free Radical Analysis

Free radicals were examined by ESR spectrometry (JES-FEIXG, JEOL, Tokyo, Japan) using manganese oxide as an internal standard. Details are as follows;

1) DPPH. 50mM of DPPH was dissolved in ethyl alcohol. One hundred microliter of this solution and 100μl of sample dissolved to distilled water or distilled water as a control were mixed for 3s then placed in an ESR spectrometry flat cell. The DPPH radicals were measured exactly after 60s. The peak intensity of DPPH of a control was set as 100%. The inhibit rate of DPPH with Manda for the peak intensity of control was shown as the rate of scavenge activity.

2) Hydroxyl Radicals. 75μl of 1mM $FeSO_4$ and 1mM DTPA, 50μl of sample or distilled water as a control, 20μl of 0.092M DMPO and 75μl of 1mM H_2O_2, were mixed for 3s in the test tube, then placed in an ESR spectrometry flat cell. The DMPO-OH spin adducts were measured exactly after 50s from putting H_2O_2. The peak intensity of DMPO-OH adducts of a control was set as 100%. The inhibit rate of DMPO-OH with Manda for the peak intensity of control was shown as the rate of scavenge activity.

3) Superoxide Radicals. Fifty microliter of 2mM HPX, 35μl of 11mM DTPA, 50μl of sample or distilled water as a control, 15μl of DMPO, and 50μl of freshly prepared XOD suspension were mixed for 3s, then placed in an ESR spectrometry flat cell. The DMPO-O_2^- spin adducts were measured exactly after just 40s. The peak intensity of DMPO-OOH adducts of a control was set as 100%. The inhibit rate of DMPO-OOH with Manda for the peak intensity of control was shown as the rate of scavenge activity.

2.3 Iron-Induced Epileptic Focus in Rats

$FeCl_2$-induced epileptic foci in rats were prepared by following procedures as previously reported by Willmore [10]. Manda were given orally to rats by a canula, 1g/kg body weight, 20 min prior to cortical injection. The rats were anesthetized by pentobarbital. Then, 5 μl of 100 mM $FeCl_2$ was injected into the left cortex at a depth of 1.2 mm. Saline was injected into the left subpial cerebral cortex of control animals with the same volume and pH as of the $FeCl_2$ treated rat. Rats were killed by decapitation 45min after the iron injection. The epileptogenic focal area was rapidly excised over an ice plate, and kept at -80°C until analysis of thiobarbituric acid reactive substances (TBARS) (Ohkawa [11]).

2.4 Statistical Analysis

Statistical analysis was performed using Student's t-test.

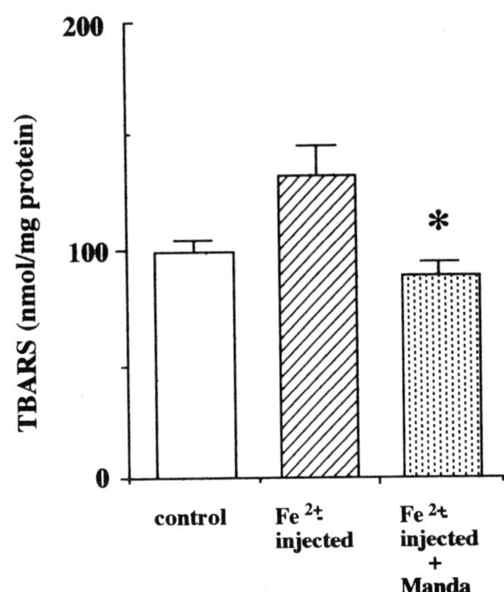

Figure 1. Effect of Manda on TBARS levels in the FeCl$_2$-induced epileptic focus of rat cerebral cortex. (means±SEM; n=5–6 determinations) * As compared with Fe^{2+}-injected group, p<0.05

3. RESULTS AND DISCUSSION

Oral administration of 1 g/kg body weight of Manda significantly inhibited the formation of TBARS, which were used as an indicator of lipid peroxidation in the FeCl$_2$-induced epileptic focus in rats (Figure 1).

Then, we examined scavenging action of Manda on DPPH, hydroxyl and superoxide radicals. We found that Manda scavenged these radicals especially superoxide radicals in a dose-dependent manner. Manda (22.7 g/L) decreased the signal of DMPO-OH in an ESR recording (41% of DMPO-OH spin adducts) (20×10^{15} spins/ml) (Figure 3). Manda decreased also the signal of DMPO-O$_2^-$ (6.8×10^{15} spins/ml) (Figure 4). Moreover Manda (25 g/L and 5 g/L) decreased DPPH radicals, respectively (Figure 2).

Figure 2. Effect of scavenging action of Manda on 1, 1-diphenyl-2-picrylhydrazyl. (means ± SEM of 3 determinations) ESR setting: magnetic field, 335.6 ± 10mT; response, 0.3sec; sweep time, 0.5min; amplitude, 1.6x1,000; modulation amplitude, 0.08mT; and power 8mW.

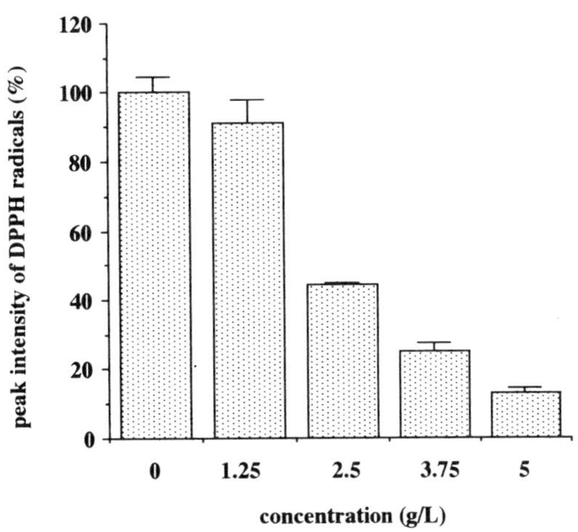

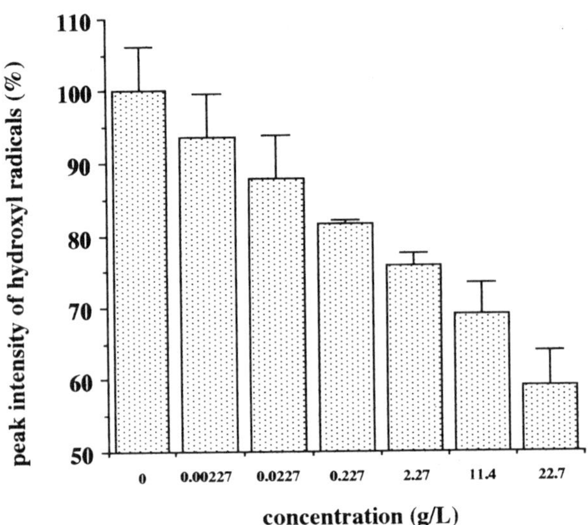

Figure 3. Effect of scavenging action of Manda on hydroxyl radicals generated from Fenton's reaction. (means±SEM of 3 determinations) ESR setting: magnetic field, 335.6±5mT; response, 0.3sec; sweeptime, 0.5min; amplitude, 1.6x1,000; modulation amplitude, 0.08mT; and power 8mW.

These results reveal that Manda is potent scavengers of superoxide radicals and have some ability in quenching hydroxyl and DPPH radicals. These radical scavenging properties may be attributed to the products by fermentation of the base plant material used in Manda preparation, though no information is available at the present.

$FeCl_2$ generates reactive oxygen radicals, in particular hydroxyl and superoxide radicals[12–15]. Both initiate and propagate peroxidation reactions [17] at the double bonds in the carbon chains of polyunsaturated fatty acids and lipids at the neuronal cell membranes. Meanwhile, antioxidants such as EPC-K1 [17, 18] and Guilingji [19] are known to inhibit

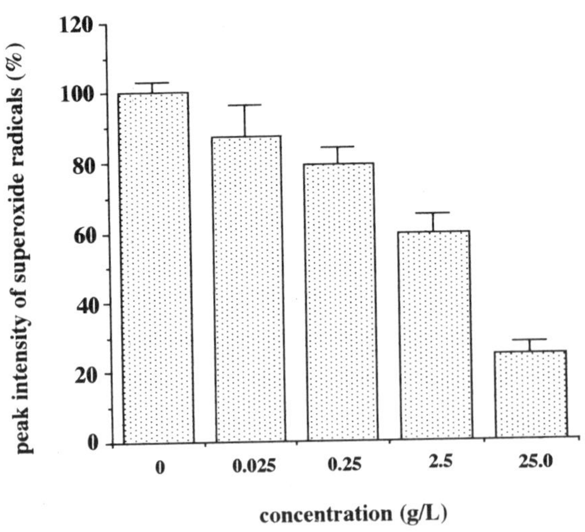

Figure 4. Effect of scavenging action of Manda on superoxide radicals generated from hypoxanthine-xanthine oxidase system. (means±SEM of 3 determinations) ESR setting: magnetic field, 335.6±5mT; response, 0.3sec; sweep time, 2min; amplitude, 1.6x1,000, modulation amplitude, 0.08mT; and power 8mW.

TBARS formation *in vivo* and prevent free radical induced cell damages. In this study, decreased TBARS formation in Manda-treated group similarly may indicate that damages by free radicals were suppressed by Manda. In conclusion, Manda scavenged free radicals, for especially superoxide radicals, and suppressed TBARS formation suggesting Manda, may be helpful to preventing neural lipid peroxidation, traumatic epilepsy, and aging [20, 21].

REFERENCES

1. Kitagawa, K., Matsumoto, M., Oda, T. et al. 1990. Neuroscience 3, 551–558
2. Braughar, J.M. 1989. Free Rad. Biol. Med., 6, 289–301
3. Oliver, C.N., Stark-Reed P.E., Stadtman, E.R., et al. 1990. Proc. Natl. Acad. Sci. USA 87, 5144–5147
4. Sun, Y. 1990. Free Rad. Biol. Med., 8, 583–599
5. Santiago, L.A., Osato, J. A., Hiramatsu, M., et al. 1991. Free Rad. Biol. Med. 11, 379–383
6. Mori, A., Hiramatsu, M., Yokoi, I. et al. 1990. Pav. J. Biol. Sci. 25, 54–62
7. Harman, D. 1981. Proc. Nat'l. Acad. Sci. USA 78, 7124–7128
8. Liu, J. and Mori, A. 1992. Neuropharmacology 31, 1287–1298
9. Kawai, M., Matsuura, S. and Mori, A. 1994. The Clinical Report. 28, 393–396
10. Willmore, L. J., Sypert G. W. and Munson J. B. 1978. Ann. Neurol. 4, 329–336
11. Ohkawa, H., Onishi, N. and Yagi, K. 1979. Analyt. Biochem. 95, 351–358
12. Rubin, J. J., Willmore, L. J., 1980. Exp. Neurol. 67, 472–480
13. Willmore, L. J., Rubin, J. J., 1981. Neurology 31, 63–69
14. Gutteridge, J. M. C. 1982. FEBS Lett., 150, 454–458
15. Willmore, L. J., Rubin, J. J., 1982. Brain Res. 246, 113–119
16. Santiago, L.A. and Mori, A. 1993. Arch. Biochem. Biophys. 306, 16–21
17. Mori, A., Hiramatsu, M., Yokoi, I., Edamatsu, R. 1990. Pav. J. Biol. Sci. 25, 54–62
18. Mori, A., Edamatsu, R., Kohno, M., Ohmori, S. 1989. Neurosciences 15, 371–376
19. Liu, J., Edamatsu, R., Kabuto, H., Mori, A. 1990. Free Radical Biol. Med. 9, 451–454
20. Yoshikawa, T, Naito, Y., Kondo, M. 1990, Neurosciences 16, 603–612
21. Mori, A., Hiramatsu, M., Hamada, H., Edamatsu, R. 1990 Neurosciences 16, 83–88

18

FREE RADICAL SCAVENGING AND ANTIOXIDANT EFFECT OF FRUCTUS MOMORDICAE

Jiankan Liu,[1] Xiaoyan Wang,[1] Muneyuki Sanada,[2] Akitoshi Natsumeda,[2] and Akitane Mori[1]

[1]Department of Neuroscience
Okayama University Medical School
[2]Sanshin & Co., Ltd.

1. INTRODUCTION

Fructus Momordicae, Luo-Han-Guo in Chinese, is the fruit of Momordica Grosvenori Swingle, that is Kuong-Guo-Mu-Bie in Chinese. This plant is perennial and is classified to cucurbitaceae, and is cultivated in such as, Yong-Fu county, Lin-Gui County, Long-Sheng County and so on, of the Guangxi Zhuangzu Autonomous Region in China.

The fruits are harvested around in September and October, after maturing, are dried by heating for one week, and then are exposed to sunshine. The dried fruits, Fructus Momordicae, are a dark brown sphere of about 5 cm in a diameter, and have been used for a special crude drug, e.g., for a febrifuge, and expectorant, or a cough medicine, among the people in China.

This fruit contains very sweet components. Some of them are "mogrosides" IV, V and VI. These are derivatives of mogrol. These are very sweet, i.e. about 400 times sweet of sucrose. The chemical structure of mogrol is 10 α-cucurbit-5-one-3β-11α, 24(R), 25-tetraol.

In the present study, we investigated the antioxidant action of Fructus Momordicae by examining the free radical scavenging and anti-peroxidative effects.

2. MATERIALS AND METHODS

One dried Fructus Momordicae, weighing about 18g, was routinely crashed by hands, heated in 1,500 ml of boiling water for 30 minutes. After the procedure, 750 ml of extract was yield. This extract contained 24.0 mg rough material of Fructus Momordicae per ml extract, which we defined as 100% extract.

Food and Free Radicals, edited by Hiramatsu *et al.*
Plenum Press, New York, 1997

Free radicals, 1,1-diphenyl-2-picrylhydrazyl, abbreviated as DPPH, superoxide and hydroxyl radicals were prepared as described elsewhere by us and were analyzed by electron spin resonance spectrometry and manganese oxide was used as an internal standard[1].

Lipid peroxidation in tissues of SD rat brain was estimated as thiobarbituric acid reactive substances as described by Ohkawa et al[2].

3. RESULTS

The figure 1 shows the effect of extract of fructus momordicae on DPPH radicals. The extract scavenged dose-dependently. IC_{50}, i.e. the concentration scavenging 50% of all administered radicals, was calculated as 30%, i.e. 0.7mg of the dried material/ml.

The figure 2 shows the effect of extract of fructus momordicae on superoxide radicals generated in the hypoxanthine-xanthine oxidase system. The extract scavenged superoxide radicals dose-dependently. IC_{50} for superoxide radicals was calculated as 21%, i.e. 5.0 mg of the dried material /ml.

Figure 3 shows the effect of extract of fructus momordicae on hydroxyl radicals generated from Fenton reagent. The extract scavenged hydroxyl radicals dose-dependently. IC_{50} was 12.5%, i.e. 3.0 mg of the dried material/ml.

Figure 4 shows the effect of extract of fructus momordicae on ferrous ion and ascorbate induced peroxidation of rat brain homogenate. The extract inhibited effectively inhibited peroxidation of rat brain homogenate dose-dependently and also time-dependently. The maximal concentration used was 10% and the inhibition rate was 40% at this concentration.

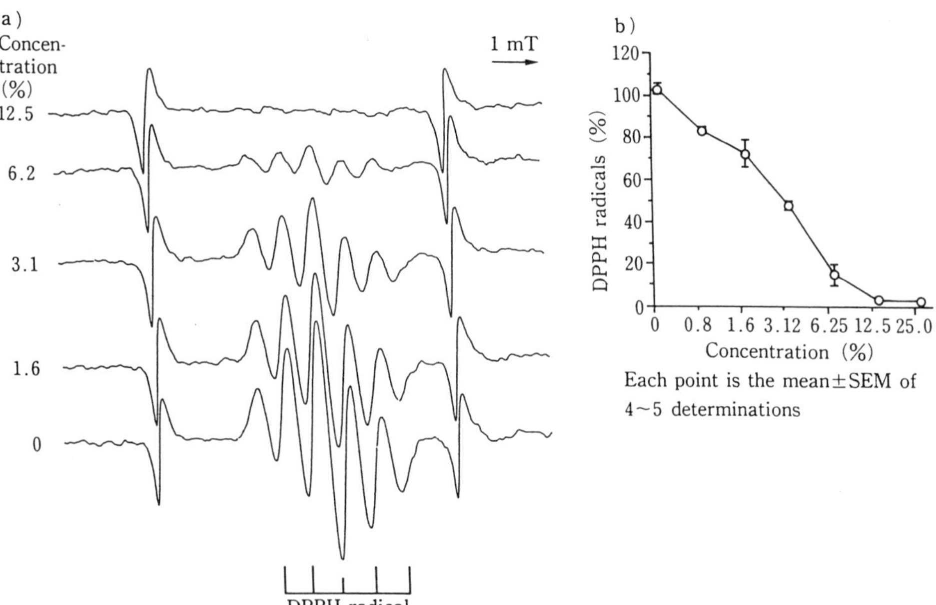

Figure 1. ESR spectra of various concentrations of extract of Fructus momordicae on DPPH radicals (a) and concentration-dependent effect (b).

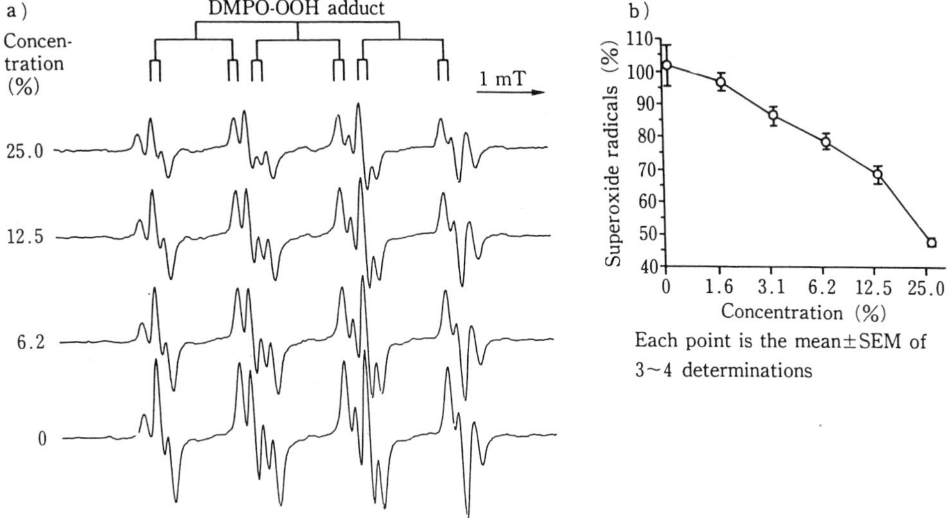

Figure 2. ESR spectra of various concentrations of extract of Fructus momordicae on superoxide radicals (a) and concentration-dependent effect (b).

4. SUMMARY

In conclusion, our electron spin resonance study showed that water extract of fructus momordicae scavenged DPPH radical and also superoxide as well as hydroxyl radical in a concentration dependent manner. Otherwise, this extract effectively inhibited the ferrous ion- and ascorbate-induced peroxidation of rat brain homogenate. These results suggest

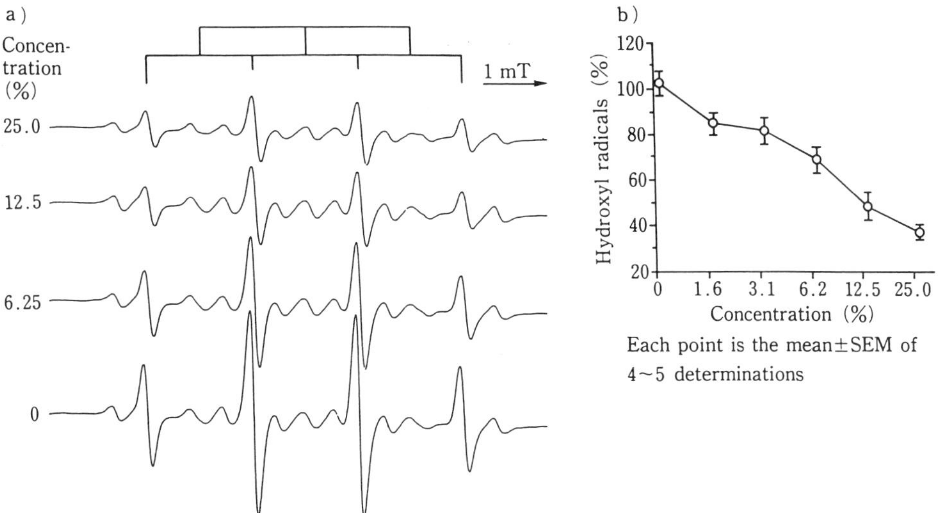

Figure 3. ESR spectra of various concentrations of extract of Fructus momordicae on hydroxide radicals (a) and concentration-dependent effect (b).

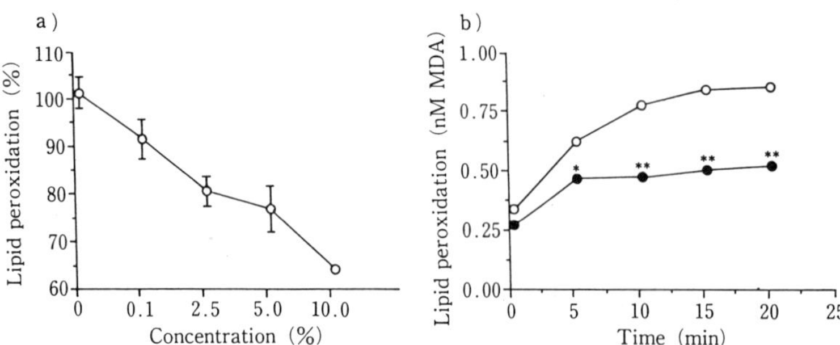

Figure 4. Effect of extract of Fructus momordicae on Fe(II)-ascorbic acid induced peroxidation in rat brain homogenate. (a) Concentration-dependent effect. Each point is the mean ± SD of 3 determinations. One hundred percent peroxidation is equal to 0.80 nM malondialdehyde. (b) Time-dependent effect. Open circle: control; solid circles: 10% extract solution was added. *: $p < 0.05$, **: $p < 0.01$ compared with control values determined by one way analysis of variance.

that one of the mechanisms of the anti-inflammatory effect of this extract may be contributive to this antioxidant action.

REFERENCES

1. Liu, J., Ogawa, N., Wang, X. and Mori, A. (1991) Free radical scavenging by fifemelane hydrochloride and its major metabolites. Arch. Int. Pharmacodyn. 311:177–187.
2. Ohkawa, H., Ohishi, N. and Yagi, K. (1979) Assay for lipid peroxides in animal tissues by thiobarbituric acid reaction. Analyt. Biochem. 95:351–358.

19

THE ROLE OF PROOXIDANT–ANTIOXIDANT IMBALANCES IN THE PATHOGENESIS OF RHEUMATOID ARTHRITIS

Erkki Antila, Tuomas Westermarck, Faik Atroshi, Visa Honkanen, and Yrjö Konttinen

Helsinki Central Institution
Kirkkonummi and Musculoskeletal Diseases and Inflammation Research
  Group
Department of Anatomy
Institute of Biomedicine
FIN-00014 University of Helsinki, Finland

1. INTRODUCTION

Free oxygen radicals seem to be involved in the pathogenesis of a variety of diseases including rheumatoid arthritis (RA)[1]. RA is primarily an autoimmune disorder where the cell mediated chronic inflammation is localized in the synovium. Free radicals are suggested to contribute to the initiation and progress of this autoimmune disease. During inflammation neutrophils and macrophages generate reduced oxygen intermediates such as superoxide radical and hydrogen peroxide. If traces of iron catalyst are present, these species can interact to form highly reactive hydroxyl-radical. Several studies using bleomycin-iron assay have shown that suitable iron catalysts for $OH^{\cdot}$ formation are present in synovial fluid taken from patients with rheumatoid arthritis. Therefore it is likely that OH is formed *in vivo*, and this may account for the degeneration of hyaluronic acid and cartilage observed in joints of rheumatoid patients[2].

Superimposed inflammatory reactions are followed by changes in antioxidant related factors which accelerate the destruction of the joint. In addition to loosely bound iron complexes, low molecular weigth loosely bound copper complexes, able to catalyze radical formation, have been detected in the synovial fluid of rheumatoid patients. Exposure of membrane lipids to oxygen radicals in the presence of iron or copper greatly enhances the process of lipid peroxidation. Indeed, as migth be expected, the rate of lipid peroxidation in rheumatoid disease is increased. Oxidant stress and hypoxic state lead to conversion of xanthine dehydrogenase enzyme to xanthine oxidase which further boosts free radical generation. Xanthine oxidase has been localized in the endothelium of synovium[3] and its inhibition reduces free radical generation from synovial tissue. Xan-

Food and Free Radicals, edited by Hiramatsu *et al.*
Plenum Press, New York, 1997

thine oxidase is elevated in the serum of rheumatic patients too[4]. Oxidant stress, due to activated xanthine oxidase and myeloperoxidase systems accelerated by intraarticular iron, degrade hyaluronate within the affected joint.

It is believed that chronic inflammation stresses protective mechanisms of the body.

It seems likely that continual production of free radicals depletes antioxidant reserves including α-tocopherol [5] and ascorbate[6]. Vitamin E and A are both lower in the serum of RA patients than in that of age matched controls[7,8].

The sulfhydryl level has been shown to be reduced in RA patients[9,10,11]. However, the glutathione level in red blood cells has been described both as low [12] and increased [14].

The total antioxidant capacity of serum has been reported to be decreased in RA compared to healthy controls[11] . Henrotin et al. [15] have suggested that the antioxidant depleting inflammatory process may be influenced by appropriate treatment with antioxidants and free radical scavengers. Reactions and treatment potentials involved, however, are highly complex. The interactions between chronic inflammation, medication and diet are still poorly known. We have tackled this problem in several studies,[7,8,15, 16, 17,18] by combining new tools, Setti-N nutritional database developed at National Health Institute in Finland to careful clinical and biochemical characterization and nutritional intervention. The results obtained were analysed with multifactorial statistical analysis.

2. TRACE ELEMENTS

Trace elements play an important role in oxygen metabolism. Free iron and copper are potential enhancers of free radical formation, other trace elements, such as zinc and selenium, protect against the harmful effects of these radicals. However, iron and copper also contribute to the antioxidant defense. It is difficult to separate the effects of trace elements on antioxidant defense and other etiological contributors.

2.1. Zinc

Low zinc found in rheumatoid arthritis patients [16] may impair T cell mediated immunity[19]. Zinc is essential for the activity of more than a hundred enzymes including delta-6-desaturase. Therefore it is not surprising that it also seems to control retinol binding protein and vitamin A level in the serum [8]. Nine independent variables such as joint score index, Waaler-Rose titre and haemoglobin predicted together 73% of the variation in serum zinc of rheumatoid arthritis patients[16] . Normally vegetable oils provide lipid soluble antioxidants and essential fatty acids. The utilization, metabolism and lipid peroxidation of fatty acids depend in addition to vitamin E on other nutrients such as zinc. A low level of zinc apparently increases oxidation of alpha-linoleic acid [20] . Interestingly we have found correlation between plasma vitamin A and serum zinc concentration. This may explained by regulation of the synthesis of retinol binding protein in liver by zinc [8]. In animal models of arthritis, zinc accumulates in liver due to induction of metallothionein. Metallothionein is regulated by IL-1, a cytokine responsible for many local and systemic inflammatory events. Infusion of recombinant IL-1 causes a decrease in serum zinc[21]

2.2. Copper

Serum copper is high in RA patients[16]. Copper indeed seems to be mobilized in inflammation. Serum-copper has the highest correlation with ceruloplasmin, which is part of the above mentioned Il-1/IL-6 mediated acute phase response[21]. The same stimulus may thus be responsible for low serum zinc and high serum copper in RA. Association between acute phase reaction and ceruloplasmin indicates homeostatic significance for ceruloplasmin because of its role as one of the major antioxidants in aqueous body humours. However, the concentration of an inflammation marker Cu-thionein is significantly diminished in patients with connective tissue diseases . Sera of patients suffering from inflammatory rheumatic diseases were almost totally depleted of this low-molecular-weight copper protein which exerts pronounced superoxide dismutase activity and scavenges effectively hydroxyl radicals and singlet oxygen. In vitro, Cu-thionein is released during oxidative burst of peripheral blood monocytes. Cortisone treatment of patients with rheumatoid arthritis replenished impressively the serum concentration of Cu-thionein [22].

Myeloperoxidase-dependent degradation of hyaluronic acid is inhibited by superoxide dismutase.[23] The activity of Cu,Zn-superoxide dismutase is dependent on copper availability.

It is also noteworthy that cytoplasmic Cu,Zn-SOD is more sensitive to the availability of copper than to that of zinc. Erythrocyte Cu,Zn-SOD acitvity is in fact increased in RA patients and it was positively correlated with serum ceruloplasmin[7]. Cu supplementation (2mg/day) for 4 weeks increased Cu,Zn-SOD activity by 21 % in erythrocytes of rheumatoid arthritis patients [24]. Furthermore, Cu found in inflamed rheumatoid synovial fluid is complexed by potent scavengers [25].

2.3. Iron

It has been known for some time that RA is accompanied by abnormalities in iron metabolism. The cytokine-influenced uptake of iron by the reticuloendothelial system leads to anaemia in RA patients. Iron is deposited within synovial cells in ferritin and haemosiderin. Inflamed synovial fluid contains elevated amounts of ferritin and low molecular mass iron as compared with normal synovial fluid [26,27]. Ferritin is an acute phase reactant and transferrin is a reverse acute phase reactant . Iron is probably initially incorporated into ferritin within macrophages which synthesize apoferritin. As levels increase it liberates iron causing oxidative damage. Free iron is taken up by fibroblast-like cells in a more stable form, hemosiderin [28]. Activated polymorphonuclear leukocytes secrete lactoferrin which is able to bind iron into a form incapable of undergoing redox reactions. Copper loaded ceruloplasmin has ferrooxidase activity contributing to antioxidase defense by inhibiting the Fe2+-catalyzed generation of hydroxyl radicals (in the presence of ascorbate).

Patients suffering from iron deficiency anemia *per se* seem to have decreased activity of essential antioxidant enzymes in their red blood cells[29]. In anaemic pregnant women, oral iron supplementation has been shown to induce a flare of rheumatoid synovitis. Thus, as in iron overload found in idiopathic or secondary haemochromatosis, iron supplementation deteriorate rheumatoid synovitis[28].

2.4. Selenium

Selenium status in rheumatoid arthritis has been recently reviewed by Tarp[30].

Selenium influences the inflammation in rheumatoid arthritis mainly through glutathione peroxidase (GSHPx). However, selenium may modulate cell-mediated immunity and favour antibody production[31] through some other mechanism. Selenium is low or normal in RA patients. This could be explained so that selenium level drops in response to inflammation. This is supported by the inverse correlation of the level of selenium in the serum and the disease duration and activity[32] and with the extent of joint inflammation[17].

This conclusion is further corroborated by observations in other inflammatory diseases such as septicaemia, pneumonia, erysipelas, and meningitis[33].

A lower level of serum selenium[30] and an increased level of plasma copper were observed in patients with rheumatoid arthritis, juvenile chronic arthritis[34] and in psoriatic arthritis in comparison with controls[35]. The activities of SOD, GSHPx and catalase have been found to be significantly decreased in patients with RA[36].

The idea to improve of rheumatoid inflammation by boosting selenium dependent antioxidant protection with supplements of selenium has not been explicitly confirmed (see 30). One point to consider is the ability of different selenium compounds to raise GSHPx activity in tissues.

2.5. Vitamin E

Vitamin E and selenium can spare one another as micronutrients in antioxidation[37]. Furthermore it has been shown that in patients with inflammatory joint disease the synovial fluid concentration of alpha-tocopherol is significantly lower relative to those of paired serum samples [5]. This indicates consumption of alpha-tocopherol within the inflamed joint via its role in terminating the process of lipid peroxidation.

3. CONCLUSIONS

The joint antioxidant defense located in various intra- and extracellular compartments protects during hypoxia-reperfusion and inflammation the functional and anatomical integrity of the joint cartilage and the viscoelastic properties of the synovial fluid. Problems arise during continuous and extensive radical generation within the joint during chronic inflammation when the essential components of the antioxidant defense become exhausted. These problems may proceed in an expanding manner through several mechanisms such as xanthine oxidase system, low molecular/ free iron catalyzed Haber-Weiss reaction and phospholipase A2 activated eicosanoid cascade. Degradation of hyaluronic acid, activation of elastase and other metalloproteinases are direct consequences of radical attack which expand tissue destruction within the joint.

Rheumatoid arthritis is not a consistent and independent disease entity: the antioxidant defense varies according to age, disease activity and duration, nutrition, digestion, drug interactions, and other diseases and allergies. To bring about clinically effective correction in the body's antioxidant protection, the following facts are crucial:

- availability of antioxidant vitamins E and C
- glutathione redox cycling: sufficient level of intracellular GSH, availability of selenium for adaptive increase of GSHPx, NADPH generation through hexosemonophosphate pathway, inhibition of GSHPx by antirheumatic drugs
- balanced fatty acid composition in membrane phospholipids for homeostatic control by eicosanoid cascades
- tissue stores and availability in nutrition of zinc, iron and copper

REFERENCES

1. Greenwald, R.A. (1991) Oxygen radicals, inflammation, and arthritis: pathophysiological considerations and implications for treatment. Semin.Arthritis.Rheum. 20: 219–240.

2. Wong, S.F., Halliwell, B., Richmond, R., Skowroneck, W.R. (1981) The role of superoxide and hydroxyl radicals in the degradation of hyaluronic acid induced by metal ions and by ascorbic acid. J.Inorg.Biochem. 14:127–132.

3. Stevens, C.R., Benboubetra, M., Harrison, R., Sahinoglu, T., Smith, E.C., Blake, D.R. (1991) Localisation of xanthine oxidase to synovial endothelium. Ann. Rheum. Dis. 50:760–762.

4. Miesel, R., Zuber, M. (1993) Elevated levels of xanthine oxidase in serum of patients with inflammatory and autoimmune rheumatic diseases. Inflammation 17:551–561.

5. Fairburn, K., Grootveld, M., Ward, R.J., Abiuka, C., Kus, M., Williams, R.B., Winyard, P.G., Blake, D.R.(1992) Alpha-tocopherol, lipids and lipoproteins in knee-joint synovial fluid and serum from patients with inflammatory joint disease. Clin. Sci. 83(6): 657–664.

6. Lunec,J., Blake, D.R. (1985) The determination of dehydroascorbic acid in the serum and synovial fluid of patients with rheumatoid arthritis (RA). Free Rad Res Comms 1:31–39.

7. Honkanen,V. (1992) , Nutritional factors in rheumatoid inflammation, Commentationes Physico-Mathematicae et Chemico-Medicae 135:1–93.

8. Honkanen, V., Konttinen, Y.T., Mussalo-Rauhamaa, H. (1989) Vitamins A and E, retinol binding protein and zinc in rheumatoid arthritis. Clin. Exp. Rheumatol. 7:465–469.

9. Lorber, A., Pearson,C.M., Meredith,W.L., Ganz-Mandell, L.E., (1964) Serum sulfhydryl determinations and significance in connective tissue diseases. Ann. Intern. Med. 61:423–434.

10. Hall, N.D., Blake, D.R. (1982) Serum SH as an indicator of phagocytic activity in rheumatoid arthritis (Abstract) Ann. Rheum. Dis. 41:00.

11. Situnayake,R.D., Thurnham,D.I., Kootathep,S., Chirico,S., Lunec,J., Davis,M., McConkey,B. (1991) Chain breaking antioxidant status in rheumatoid arthritis: clinical and laboratory correlates. Ann.Rheum.Dis. 50: 81–86.

12. Banford, J.C., Brown, D.H., Hazelton, R.A., McNeil, C.J., Smith, W.E., Sturrock, R.D. (1982) Altered thiol status in patients with rheumatoid arthritis. Rheumatol. Int 2:107–111.

13. Koster, J.F.., Biemond, P., Swaak, A.J.G. (1986) Intracellular and extracellular sulphydryl levels in rheumatoid arthritis. Ann. Rheum. Dis. 45:44–46.

14. Munthe, E., Kåss, E., Jellum, E. (1981) D-penicillamine-induced increase in intracellular glutathione correlating to clinical response in rheumatoid arthritis. J. Rheumatol. 8 (Suppl. 7): 14–19.

15. Henrotin,Y., Deby-Dupont, G., Deby, C., Franchimont, P., Emerit, I. (1992) Active oxygen species, articular inflammation and cartilage damage, EXS. 62: 308–322.

16. Mussalo-Rauhamaa, H., Konttinen, Y.T., Lehto, J., Honkanen, V. (1988) Predictive clinical and laboratory parameters for serum zinc and copper in rheumatoid arthritis. Ann. Rheum. Dis. 47:816–819.

17. Honkanen, V., Lamberg-Allardt, C., Lehto, J., Vesterinen, M., Mussalo-Rauhamaa, H., Metsä-Ketelä, T., Westermarck, T., Konttinen, Y.T. (1991) Plasma trace element levels in rheumatoid arthritis, the influence of dietary factors and disease activity. Am.J. Clin. Nutr. 54:1082–1086.

18. Nordstrom, D.C.E., Honkanen, V.E.A., Nasu, Y., Antila, E., Friman, C., Konttinen, Y.T. (1995) Alpha-linolenic acid in the treatment of rheumatoid arthritis - Adouble-blind, placebo- controlled and randomized study - flaxseed vs safflower seed. Rheumatology International. 14(6):231–234.

19. Good, R.A., Fernandes, G., Garofalo, J.A., Cunningham-Rundles, C., Iwata, T., Wes, A. (1982) Zinc and immunity. In: A.S. Prasad (ed), Clinical, biochemical and nutritional aspects of trace elements, pp. 189–202. Alan R. Liss Inc., New York.

20. Cunnane, S.C., Yang, J., Chen, Z.Y. (1993), Low zinc intake increases apparent oxidation of linoleic and α-linolenic acids in the pregnant rat. Can. J. Physiol. Pharmacol. 71:205–210.

21. Cousins, R.J. (1985) Absorption, transport, and hepatic metabolism of copper: special reference to metallothionein and ceruloplasmin. Physiol. Rev. 65: 238- 309.

22. Miesel,R., Zuber,M. (1993) Copper-dependent antioxidase defenses in inflammatory and autoimmune rheumatic diseases. Inflammation 17(3): 283–294.

23. Lindvall, S., Rydell, G. (1994) Influence of various compounds on the degradation of hyaluronic acid by a myeloperoxidase system. Chem. Biol.Interact. 90: 1–12.

24. DiSilvestro, R.A., Marten, J., Skehan, M. (1992) Effects of copper supplementation on ceruloplasmin and copper-zinc superoxide dismutase in free-living rheumatoid arthritis patients. J. Am. Coll. Nutr. 11: 177–180.

25. Naughton, D.P., Knappitt, J., Fairburn, K., Gaffney, K., Blake, D.R., Grootveld, M. (1995) Detection and investigation of the molecular nature of low-molecular-mass copper ions in isolated rheumatoid knee-joint synovial fluid. FEBS-Lett. 361: 167–172.

26. Mowat, A.G., Hothershall, T.E. (1968) Nature of anaemia in rheumatoid arthritis. VIII Iron content of synovial tissue in patients with rheumatoid arthritis and in normal individuals. Ann. Rheum. Dis. 27: 345–350.

27. Morris, C.J., Blake, D.R., Wainwrigth, A.C., Steven, M.M.(1982) Relationship between iron deposits and tissue damage in the synovium: an ultrastuctural study. Ann. Rheum.Dis. 45:21.

28. Morris, C.J., Earl, J.R., Trenam, C.W., Blake,D.R. (1995) Reactive oxygen species and iron—a dangerous partnership in inflammation. Int. J. Biochem. Cell. Biol. 27: 109–122.

29. Bartal, M., Mazor, D., Dvilansky, A., Meyerstein, N. (1993) Iron deficiency anemia: recovery from in vitro oxidative stress. Acta Haematol. 90: 94–98.

30. Tarp, U. (1994) Selenium and the selenium-dependent glutathione peroxidase in rheumatoid arthritis. Danish Medical Bulletin 41:264–274.

31. Koller, L.D., Exon, J.H., Talcott, P.A., Osborne, C.A., Henningsen, G.M., (1986) Immune responses in rats supplemented with selenium. Clin. Exp. Immunol. 63: 570–576.

32. Tarp U., Overvad, K., Hansen, J.C., Thorling, E.B. (1985) Low selenium level in severe rheumatoid arthritis. Scand. J. Rheumatol. 14:97–101.

33. Srinivas, U., Braconier, J.H., Jeppsson, B., Abdulla, M., Åkesson, B., Öckerman, P.A. (1988) Trace element alterations in patients with reduced selenium state. Eur. J. Pediatr. 48:495–500.

34. Honkanen, V., Pelkonen, P., Mussalo-Rauhamaa, H., Lehto, J., Westermarck, T. (1989) Serum trace elements in juvenile chronic arthritis. Clin. Rheumatol 8:64–70.

35. Azzini, M., Girelli, D., Olivieri, O., Guarini, P., Stanzial, A.M., Frigo, A., Milanino, R., Bambara, L.M., Corrocher, R. (1995). Fatty acids and antioxidant micronutrients in psoriatic arthritis. J Rheumatol 22(1):103–108.

36. Imadaya, A., Terasawa, K., Tosa,H., Okamoto, M., Toriizuka,K. (1988) Erythrocyte antioxidant enzymes are reduced in patients with rheumatoid arthritis. Rheumatol 15:1628–1631.

37. Levander, O.A. (1982) Clinical consenquences of low selenium intake and its relationship to vitamin E. Ann. NY Acad. Sci. 355:70–82.

20

FREE RADICALS AND ANTIOXIDANTS IN TRICHOTHECENS TOXICITY

Faik Atroshi,[1] Aldo Rizo,[2] Erkki Antila,[1] and Tuomas Westermarck[3]

[1]Department of Clinical Sciences
Vet. Medicine
Dept. of Anatomy, University of Helsinki
[2]National Vet. Institute
Helsinki
[3]Helsinki Central Institution
Killinmäki, Kirkkonummi
Finland

1. INTRODUCTION

Antioxidants are of primary value biologically in restricting the damage that reactive free radicals can cause to cells and cellular components. Clearly, antioxidants are much more likely to be effective against chemicals that, to a significant extent, are metabolically activated to reactive free radical intermediates, than against compounds whose metabolic activation results in nonradical intermediates such as epoxides or carbon ions. The antioxidant defense mechanisms, by which the body protects itself against free radical damage, consists of low molecular weight free radical scavengers (α-tocopherol, ascorbate, 13-carotene/ carotenoids) and of enzyme systems such as superoxide dismutase (SOD), catalase (CAT). During oxidative stress, hydroperoxides are the only species that may accumulate at rather high steady-state concentrations. The main pathways available for degradation of such hydroperoxides are that of glutathione peroxidase, whose biologically significant members probably contain selenium at their active site.

Mycotoxins are a divers group of relatively small molecules, produced by fungi that elicite a wide range of toxic responses in animals and humans. To assess the physiological importance of this reductive pathway, we looked at the protective effect of natural antioxidants (vitamins A, C and E and selenium) on T-2 toxin and deoxynivalenol (DON) toxicity in the liver of rats as part of a comprehensive effort to curtail the adverse health effects posed by these mycotoxins.

Food and Free Radicals, edited by Hiramatsu *et al.*
Plenum Press, New York, 1997

2. MATERIAL AND METHODS

The experiment was conducted on 80 Wistar male rats with body weight of 160–180 g. The animals were divided into ten groups of eight rats each. For two weeks the rats were fed two different semi-synthetic feeds. After two weeks on the diet, the rats of groups 2, 9, 10 and groups 3, 7, 8 received orally a single dose of DON (18 mg/kg b.w.) or T-2 toxin (2.8 mg/kg b.w.) respectively, in 2% ethanol. The control groups 1, 4, 5, 6 received 2% ethanol and no toxins. Twenty-four hours after the treatment with mycotoxins, all rats were decapitated. Enzyme acitvity were analyzed according to the method descibed earlier[1].

3. RESULTS AND DISCUSSION

Rats of group 1 (control deficient feed, Table 1), had 7% higher TBA values than those of group 4 (control conventional feed). Group 4 rats showed 22% higher TBA values than group 6 ($p<0.05$), which received vitamin A, C, E and selenium. Relative to group 4, there was no increase in the MDA value in groups 9 and 10, which received DON with selenium and vitamins (A, C and E), but increases of 78% ($p<0.01$) and 57% ($p<0.01$) were recorded for groups 7 and 8, which received T-2 toxin with vitamins and selenium. However, evaluated against their own controls, groups 2 and 3 (deficient feed) which were treated with DON and T-2 toxin produced the highest TBA values. Groups 5 and 6, which were given selenium plus vitamins, respectively, had TBA values lower than the control (group 4).

Toxic effects of T-2 toxin and DON are either acute or chronic, depending in large part on the dose and duration of exposure. However, because many of the sequelae are not specific for T-2 or DON, field diagnosis often are difficult. Investigators studying the mechanisms of toxicity thus far have focused on the effects of these mycotoxins on protein and macromolecular synthesis, membrane function, and immune parameters (Figure 1). A number of trichothecene mycotoxins inhibited protein synthesis in eukaryotic cells[2]. The increase in lipid peroxide and the decrease in CAT, SOD, cytochrome P-450, and GSH-transferase activities in our earlier study[1], as the result of T-2 toxin and DON administration to rats fed a deficient diet are of particular interest. Free radicals involvement, and the scavenging of these radicals by CAT and SOD, are possible. Oral adminstration of

Table 1. TBA values in the livers of rats treated with mycotoxins

Group no.	mg MDA/g liver (average value)	Toxin	Type of feed
1	0.30	—	Deficient (control)
2	0.40	DON	
3	0.92	T-2	
4	0.28	—	Conventional (control)
5	0.24	—	+ Se
6	0.23	—	+Se+vitamins
7	0.50	T-2	+ Se
8	0.44	T-2	+Se+vitamins
9	0.24	DON	+ Se
10	0.22	DON	+Se+vitamins

T-2 toxin and DON to rats in our study stimulated lipid peroxidation in rat liver. Lipid peroxidation is considered both a pathological and toxicological process in biological systems. It appears to be involved in vitamin E deficiency and ethanol poisoning, and many other degenerative diseases. Lipid peroxides, produced by the oxidation of lipids by O_2^- and H_2O_2 will also affect mitochondrial functions. Linoleat hydroperoxide is an effective uncoupler of oxidative phosphorylation, lipid peroxides also affect mitochondrial membrane permeability to cations[3]. The oxidation of lipids, mediated by O_2^- production, and the reduction of lipid peroxides by the glutathione system can therefore serve as a regulatory system of mitochondrial function even if the NAD(P)H/NAD(P) ratio is itself no significantly altered. The peroxidation of polyunsaturated fatty acids present in membranes has been proposed as a mechanism responsible for the toxicity of many xenobiotics and toxins[4]. The toxicity of these compounds is related to the intracellular glutathione levels. Even though the intracellular concentration is the millimolar range, depletion of this intracellular thiol leads to increased cellular damage[5]. Microsomal epoxide hydrolase plays an important role in the metabolic detoxication of trichothecenes in rats. In contrast to this, however, the contribution of hepatic epoxide hydrolase to the transformation of trichothecenes is quite small and that most the de-epoxytrichothecenes are produced by strict anaerobs of the intestinal microflora of rats[6]. In in vitro experiments conducted with T-2 toxin on mice, rat and monkey liver, they did not find any de-epoxytrichothecenes'. In

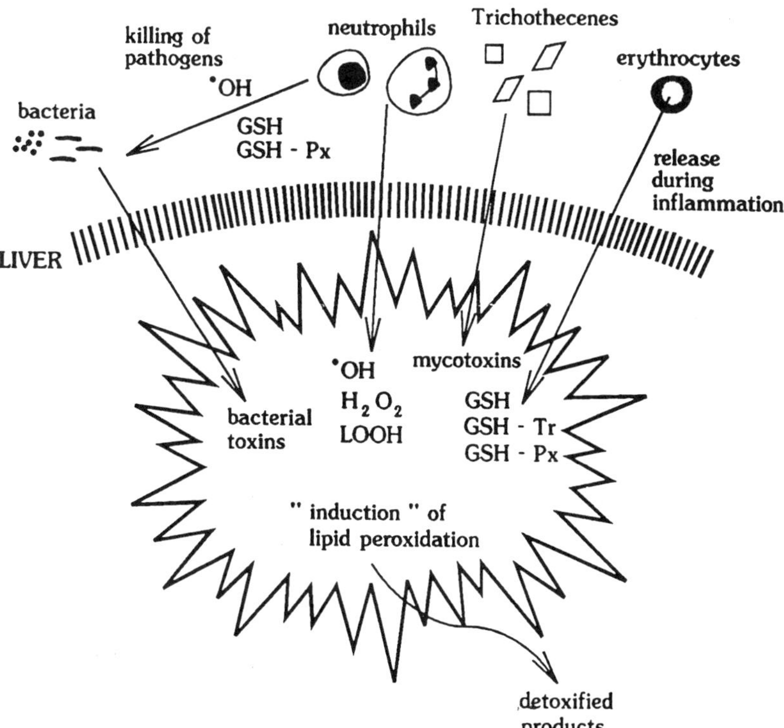

Figure 1. A multistep toxicity model. Possible effects of trichothecenes on the intracellular GSH metabolism. GSH represents the intracellular charged compounds, a potential source of reducing equivalents. The intracellular GSH and GSH-Px may greatly be affected by oxygen radicals and lipid peroxidation, associated e.g. in toxicity and hypoxia.

conclusion, the avialable evidence indicates that selenium, in combination with ascorbic acid and α-tocopherol may counteract the toxicity of trichothecenes such as T-2 toxin and DON which exert their toxic effects via oxygen radical production. Free radicals often contribute to toxic mechanism. In order for maximum protection to occur, selenium, vitamin A, vitamin E and vitamin C must be given either shortly before or during the exposure to the trichothecenes. Antioxidant supplementation might thus be worth testing in an attempt to counteract trichothecenes toxicity.

REFERENCES

1. Rizo, A.F., Atroshi, F., Ahotupa, M., Sankari, S., and Elovaara, R.A.(1994) Protective effect of antioxidants against free radical-mediated lipid peroxidation induced by DON or T-2 toxin. J. Vet. Med. A,41:81–90.
2. Ueno, Y., Nakajima, M., Sakai, K., Ishii, K., Sato, N. and Shimada, N.(1973) Comparative toxicology of trichothecene mycotoxins: Inhibition of protein synthesis in animal cells. J.Biochem.74:285–296.
3. Forman, H.J. and Boveris, A. (1982) Superoxide radical and hydrogen peroxide in mitochondria.In: W. Pryor(ed), Free Radical in Biology, pp.65–90. Academic Press.
4. Tsuchida, M., Miura, T., Shimizu, T. and Aibara, K.(1984) Elevation of thiobarbituric acid values in the rat liver intoxicated by T-toxin. Biochem Med. 31:147–166.
5. Tutelyan, V.A., Kravchenko, L.V., Kuzmina, E.E., Avrenieva, L.l. and Kumpulainen, J.T.(1990) Dietary selenium protects against acute toxicity of T-2 toxin in rats. Food Add.Contamin 7:821–827.
6. Yoshizawa, T., Okamoto, K., Sasamoto, T. and Kuwamura, K.(1985) In vivo metabolism of T-2 toxin, a trichothecene mycotoxin. On the formation of depoxydation products.Proc. Jpn. Assoc.Mycotoxicol. 21, 9–12,1985.

21

LONG-TERM FOLLOW-UP OF TWO DUCHENNE MUSCLE DYSTROPHY PATIENTS TREATED WITH ANTIOXIDANTS

Tuomas Westermarck,[1] Erkki Antila,[2] Satu Kaksonen,[3] Juha Laakso,[4] Matti Härkönen,[5] and Faik Atroshi[6]

[1]Helsinki Central Institution
Killinmäki, 02400 Kirkkonummi
[2]Department of Anatomy
University of Helsinki
[3]Ruskeasuo School
[4]Mineral Lab. Mila, Ltd
[5]Meilahti Hospital
[6]Faculty of Vet. Medicine
Helsinki, Finland

1. INTRODUCTION

Duchenne muscular dystrophy (DMD) is one of the most severe frequent myopathies, which evolves progressively until it causes disablement and death, around the age of twenty. It is recessively inherited, sex-linked, and the gene determining DMD is Xp-21 located. Its incidence is 1/3500 of male births and one third of the cases is originated from a new mutation. The defected gene causes a shortage or absence of a structural protein dystrophin, which is near the sites of Ca^{2+} release from sarcoplasmic reticulum and uptake of intracellular Ca^{2+}. The genetic alteration produces an abnormality in the membrane of muscular fibers that consists of a disturbance in the calcium transport (Ca^{++}), inside the muscular fibers, which is the base mechanism of cellular degeneration and necrosis.

A nucleotide degradation, and decreased muscle ATP and ADP content has been reported.

Therapy of DMD has been an elusive goal. Studies with isolated myocytes have shown that lipid peroxidation with an enhanced free radical production can be activated by increasing Ca concentration. No wonder that several kinds of antioxidants have been proposed as a treatment since increased levels of thiobarbituric acid (TBA) reactive material has been found in the muscles and blood of patients with DMD.

Vitamin E has been observed to decrease the amount of TBA reactive material in dystrophic muscle. Previously we have reported that the biological half-life of

Food and Free Radicals, edited by Hiramatsu *et al.*
Plenum Press, New York, 1997

selenium-75 (^{75}Se) in DMD patients is significantly shorter than in healthy controls.[1] On the basis of these facts we started to treat some DMD patients with selenium and other antioxidants[2].

2. MATERIALS AND METHODS

In 1981 the antioxidant treatment of two siblings with DMD, aged six and ten years, was started. At that time the elder brother was wheel-chair bound and was not able to walk since eight years of age, the younger one was still able to walk almost normally.

Selenium (Se) has been given as sodium selenite, 0.1 mg Se kg^{-1} b.w. day^{-1}; α-tocopherol 10–20 mg.kg^{-1} b.w.day^{-1}; vitamin B$_2$, 0.2 mg.kg^{-1} b.w day^{-1}; vitamin B$_6$, 5 mg.kg^{-1} b.w. day^{-1}. For three years he has been administered ubiquinol-10 (coenzyme Q$_{10}$) and carnitine. The levels of vitamin E, coenzyme Q$_{10}$, and selenium in serum and eryhtrocyte GSH-Px were analysed according to the methods described earlier[1].

3. RESULTS AND DISCUSSION

Normally patients with DMD become wheel-chair bound between the ages of 8 and 12 years. The condition of the elder brother, who was wheel-chair bound at the age of 8 years, was gradually deteriorated. He was not able to move himself at all when he deceased at 17 years of age. However, at the age of 15 years the younger brother was still able to walk without any assistance. Six months later he became wheel-chair bound. His weight is 55 kg and his height is about 160cm. The present daily megadoses of nutrient supplements (vitamin E 1200 mg; sodium selenite 8 mg; riboflavin 3 mg; pyridoxine 75 mg; carnitine 600 mg; and coenzyme Q$_{10}$ 180 mg) have been well tolerated and no side effects have been observed. Now this 18-year old DMD youth is still daily able to swim for 45 min without any assistance and he is one of the most talented in his school and will be graduated next year. The mean IQ of DMD patients is without treatment 15–20 points lower than that of the normal population, and about one-third of all patients have a significant but non-progressive mental handicap[3].

This long-term pilot study speaks in favor for a larger study. Dr Anders Erikson from Sweden (personal communication) has made the same observations with two Swed-

Table 1. Total and free serum carnitine (μmol/l) in DM patients aged 7–18 years

	Total	Free
Untreated DMD-patients		
HJ	58.9	32.9
TJ	50.1	29.6
HP	87.2	70.6
KJ	69.1	48.7
RK	68.1	49.6
Treated with carnitine and coenzyme Q$_{10}$		
PJ	53.5	37.0
JJ	66.6	43.5

ish siblings suffering from DMD, as we have also made with two 13- and 14-year old cousins with DMD.

Mice with genetic muscular dystrophy have been observed to absorb and retain larger amounts of total Se than paired control, however, Westermarck et al.[1] showed that humans with genetic muscular dystrophy did not differ in this respect from healthy controls. Survival time of dystrophic mice treated with an strong antioxidant (0.03 % LY256548) has been reported to increase significantly, and LY 178002 doubled the life span of dystrophic mice[4].

REFERENCES

1. Westermarck, T., Rahola, T., Kallio,A.K., Suomela, M. (1982) Long term turnover of selenite-Se in children with motor disorders. Klin. Pädiat.194:301–302.
2. Jackson, M.J., and Edwards, R.H.T.(1990) Free radicals and trials of antioxidant therapy in muscle diseases. Adv. Exp. Med. Biol. 264:485–91.
3. Hyser, C.L. and Mendell, J.R.(1988) Recent advances in Duchenne and Becker muscular dystrophy.Neurol. Clin. 6:429–453.
4. Panetta, J.L., Philips, M.L., Benslay, D.N., Ho, P.K., Towner, R., Roush, M.E. and Clements J.(1990) Two anti-oxidants, LY 178002 and LY 256548, as therapies for three unrelated degenerative diseases.Free Radical Biol. Med.9, suppl.1:191.